Springer Proceedings in Physics 78

Springer Proceedings in Physics

Managing Editor: H. K. V. Lotsch

46 *Cellular Automata and Modeling of Complex Physical Systems* Editors: P. Manneville, N. Boccara, G. Y. Vichniac, and R. Bidaux
47 *Number Theory and Physics* Editors: J.-M. Luck, P. Moussa, and M. Waldschmidt
48 *Many-Atom Interactions in Solids* Editors: R. M. Nieminen, M. J. Puska, and M. J. Manninen
49 *Ultrafast Phenomena in Spectroscopy* Editors: E. Klose and B. Wilhelmi
50 *Magnetic Properties of Low-Dimensional Systems II: New Developments* Editors: L. M. Falicov, F. Mejía-Lira, and J. L. Morán-López
51 *The Physics and Chemistry of Organic Superconductors* Editors: G. Saito and S. Kagoshima
52 *Dynamics and Patterns in Complex Fluids: New Aspects of the Physics–Chemistry Interface* Editors: A. Onuki and K. Kawasaki
53 *Computer Simulation Studies in Condensed-Matter Physics III* Editors: D. P. Landau, K. K. Mon, and H.-B. Schüttler
54 *Polycrystalline Semiconductors II* Editors: J. H. Werner and H. P. Strunk
55 *Nonlinear Dynamics and Quantum Phenomena in Optical Systems* Editors: R. Vilaseca and R. Corbalán
56 *Amorphous and Crystalline Silicon Carbide III, and Other Group IV-IV Materials* Editors: G. L. Harris, M. G. Spencer, and C. Y. Yang
57 *Evolutionary Trends in the Physical Sciences* Editors: M. Suzuki and R. Kubo
58 *New Trends in Nuclear Collective Dynamics* Editors: Y. Abe, H. Horiuchi, and K. Matsuyanagi
59 *Exotic Atoms in Condensed Matter* Editors: G. Benedek and H. Schneuwly
60 *The Physics and Chemistry of Oxide Superconductors* Editors: Y. Iye and H. Yasuoka
61 *Surface X-Ray and Neutron Scattering* Editors: H. Zabel and I. K. Robinson
62 *Surface Science: Lectures on Basic Concepts and Applications* Editors: F. A. Ponce and M. Cardona
63 *Coherent Raman Spectroscopy: Recent Advances* Editors: G. Marowsky and V. V. Smirnov
64 *Superconducting Devices and Their Applications* Editors: H. Koch and H. Lübbig
65 *Present and Future of High-Energy Physics* Editors. K.-I. Aoki and M. Kobayashi
66 *The Structure and Conformation of Amphiphilic Membranes* Editors: R. Lipowsky, D. Richter, and K. Kremer
67 *Nonlinearity with Disorder* Editors: F. Abdullaev, A. R. Bishop, and S. Pnevmatikos
68 *Time-Resolved Vibrational Spectroscopy V* Editor: H. Takahashi
69 *Evolution of Dynamical Structures in Complex Systems* Editors: R. Friedrich and A. Wunderlin
70 *Computational Approaches in Condensed-Matter Physics* Editors: S. Miyashita, M. Imada, and H. Takayama
71 *Amorphous and Crystalline Silicon Carbide IV* Editors: C. Y. Yang, M. M. Rahman, and G. L. Harris
72 *Computer Simulation Studies in Condensed-Matter Physics IV* Editors: D. P. Landau, K. K. Mon, and H.-B. Schüttler
73 *Surface Science: Principles and Applications* Editors: R. F. Howe, R. N. Lamb, and K. Wandelt
74 *Time-Resolved Vibrational Spectroscopy VI* Editors: A. Lau, F. Siebert, and W. Werncke
75 *Computer Simulation Studies in Condensed-Matter Physics V* Editors: D. P. Landau, K. K. Mon, and H.-B. Schüttler
76 *Computer Simulation Studies in Condensed-Matter Physics VI* Editors: D. P. Landau, K. K. Mon, and H.-B. Schüttler
77 *Quantum Optics VI* Editors: D. F. Walls and J. D. Harvey
78 *Computer Simulation Studies in Condensed-Matter Physics VII* Editors: D. P. Landau, K. K. Mon, and H.-B. Schüttler

Volumes 1–45 are listed at the end of the book

D. P. Landau K. K. Mon H.-B. Schüttler (Eds.)

Computer Simulation Studies in Condensed-Matter Physics VII

Proceedings of the Seventh Workshop
Athens, GA, USA, 28 February – 4 March 1994

With 85 Figures

Springer-Verlag
Berlin Heidelberg New York
London Paris Tokyo
Hong Kong Barcelona
Budapest

Professor David P. Landau, Ph. D.
Professor K. K. Mon, Ph. D.
Professor Heinz-Bernd Schüttler, Ph. D.

Center for Simulational Physics
The University of Georgia
Athens, GA 30602, USA

ISBN-13: 978-3-642-79295-3 e-ISBN-13: 978-3-642-79293-9
DOI: 10.1007/ 978-3-642-79293-9

CIP data applied for

Softcover reprint of the hardcover 1st edition 1994

Typesetting: Camera ready copy from the authors/editors
SPIN: 10469612 54/3140 - 5 4 3 2 1 0 - Printed on acid-free paper

Preface

Computer simulations are playing an expansive role in both research and instruction in many areas of physics. As the simulations community began to grow, we sensed a need for a meeting place for both experienced simulators and neophytes to discuss new techniques and recent results in an environment which promoted extended discussion. As a consequence, the Center for Simulational Physics established an annual workshop on Recent Developments in Computer Simulation Studies in Condensed-Matter Physics. This year's workshop was the seventh in this series, and the interest shown by the scientific community demonstrates quite clearly the useful purpose which the series has served. The latest workshop was held at The University of Georgia, February 28 to March 4, 1994, and these proceedings form a record which is published with the goal of timely dissemination of the material to a wider audience.

These proceedings are composed of four parts. The first part contains invited papers which deal with simulational studies of classical systems and includes an introduction to some new simulation techniques. The second part of the proceedings is devoted to invited papers on quantum systems including new results for strongly correlated electron and quantum-spin models. The third part contains a single paper which considers the question of computer security, a topic of ever-increasing importance. The final part is comprised of contributed presentations.

We hope that all our readers will benefit from papers in their own speciality as well as profit from exposure to new algorithms, methods of analysis, and conceptual developments. We have already learned that fruitful collaborations and new research projects have resulted from these gatherings at The University of Georgia, and we hope that these proceedings will further expand this tendency.

This year's workshop was made possible through the generous support of the National Science Foundation and the Center for Simulational Physics at The University of Georgia.

Athens, GA
June 1994

D.P. Landau
K.K. Mon
H.-B. Schüttler

Contents

Computer Simulation Studies in Condensed-Matter Physics: An Introduction
By D.P. Landau, K.K. Mon, and H.-B. Schüttler 1

Part I **Classical Systems**

Complexity in Rayleigh-Bénard Convection
By H.W. Xi, J.D. Gunton, and J. Viñals (With 5 Figures) 7

Numerical Zero-Temperature Results for the 3*d* Edwards-Anderson Ising Spin Glass
By B.A. Berg (With 7 Figures) 18

Recent Developments in Monte-Carlo Simulations of First-Order Phase Transitions
By W. Janke (With 6 Figures) 29

Statistical Dependence and Related Topics
By M. Kikuchi, N. Ito, and Y. Okabe (With 7 Figures) 44

Dynamical Scaling of the Island-Size Distribution and Percolation in Submonolayer Molecular Beam Epitaxy
By J.G. Amar, F. Family, and P.M. Lam (With 5 Figures) 57

Computer Simulation of the Structure Factor
By P.A. Lindgård (With 7 Figures) 69

Part II **Quantum Systems**

The Density-Matrix Renormalization Group for Fermion Systems
By R.M. Noack, S.R. White, and D.J. Scalapino (With 7 Figures) 85

Simulation of the Electron-Phonon Interaction in Infinite Dimensions
By J.K. Freericks and M. Jarrell (With 3 Figures) 99

Excitation Spectra in the 1-D Hubbard Model from Quantum-Monte-Carlo Simulations
By A. Muramatsu, R. Preuss, W. von der Linden, P. Dieterich, F.F. Assaad, and W. Hanke (With 8 Figures) 113

Simulating $2d$ Fermions in Strong Magnetic Fields
By G. Ortiz (With 4 Figures) 125

Part III Computer Security

Computer Security in Scientific and Administrative Environments
By R.W. Gerling (With 4 Figures) 145

Part IV Contributed Papers

Monte Carlo with Absorbing Markov Chains:
A Faster Monte-Carlo Algorithm for Dynamical Studies
By M.A. Novotny (With 2 Figures) 161

Non-Arrhenius Conductivity
in a Driven System of Interacting Lattice Gas
By R.B. Pandey (With 4 Figures) 166

Non-Equilibrium Steady States of Ising Systems
By K.E. Bassler (With 3 Figures) 171

Fourier-Accelerated Polymer Dynamics
By B. Dünweg (With 1 Figure) 177

A Multicanonical Study of Non-Polar Amino Acids
By U.H.E. Hansmann and Y. Okamoto (With 3 Figures) 183

Classical and Quantum Two-Dimensional Fluids in the Gibbs Ensemble
By F. Schneider, M.-O. Ihm, and P. Nielaba 188

Optimization by Multicanonical Annealing
and the Traveling-Salesman Problem
By J. Lee and M.Y. Choi (With 1 Figure) 193

Numerical Studies of Metastability in Ising Systems:
Short-Range Versus Long-Range Interactions
By B.M. Gorman and C.C.A. Günther (With 2 Figures) 199

Study of Frustrated Systems Via Dissipative Maps
By G.S. Canright and G.I. Watson (With 1 Figure) 204

Representation of Relaxation Phenomena by Lattice-Gas Models
By P. Lallemand 209

Some Monte-Carlo Studies of the Ising Model
By G.A. Baker, Jr. and J.J. Erpenbeck (With 1 Figure) 213

Decoupled-Cell Quantum Monte Carlo:
How Should We Calculate the Energy?
By C.J. Sisson (With 2 Figures) 219

Computer Study of a $d=2$ Mixed Ising Ferrimagnet
By G.M. Buendıa, M.A. Novotny, and J. Zhang (With 2 Figures) 223

Computational Studies of Organolithium Compounds
By L.M. Pratt and I.M. Khan 228

Index of Contributors 233

Computer Simulation Studies in Condensed-Matter Physics: An Introduction

D.P. Landau, K.K. Mon, and H.-B. Schüttler

Center for Simulational Physics, The University of Georgia,
Athens, GA 30602-2451, USA

The 7th Annual Workshop featured presentations dealing with several different facets of the rapidly developing area of computer simulations in condensed matter physics. This year's proceedings are divided into four parts. Parts I and II contain papers resulting from the invited presentations, and are rather long papers having some pedagogical content. Part III is a special section dealing with the question of data security. Part IV is the final section of this volume in which a number of shorter, contributed papers focusing on very recent research results are presented.

The opening paper by H.-W. Xi et al. discusses complexity in Rayleigh-Benard convection with a generalized Swift-Hohenberg Model. Numerical solutions show for the first time spatiotemporal spiral pattern formation in good qualitative agreement with experimental studies. B. A. Berg then presents numerical zero temperature results for the three-dimensional Edwards-Anderson Ising spin glass model. Finite-size scaling analysis of the data is used to compare with two different theoretical scenarios: Parisi mean field theory versus a droplet scaling ansatz. For the lattice sizes considered, the droplet scaling ansatz is found to be consistent with the data. The author points out that this does not rule out the mean field theory completely as much larger lattices are needed to be conclusive. In the following paper, W. Janke discusses progress in both algorithms and means of analysis for systems undergoing first order transitions. He describes a multicanonical-multigrid Monte Carlo method and shows how improved accuracy in the finite size analysis can be obtained with the inclusion of several new observables. M. Kikuchi et al. consider the relation of the equilibrium relaxation time, the susceptibility and the statistical error for data from Monte Carlo simulations. A quantity called the statistical dependence time is introduced and a method for calculating equilibrium relaxation time is proposed. The authors apply the method to obtain the dynamical critical exponent for the Ising model in two and three dimensions. J. G. Amar et al. then present a detailed study of simulations of fim growth by molecular beam epitaxy in the submonolayer regime. They use a model appropriate for dendritic island growth, and from an analysis carried out for several properties they discover four different scaling regimes. P.-A. Lindgård points out the recent interest and importance of the structure factor. A number of examples in which computer simulation has been very useful in understanding the structure observed by scattering techniques are discussed. They include lattice gas models for high T_c superconductors, adsorbed particles on graphite, phase separation and spinodal decomposition.

In the first paper in the section on quantum systems, R. M. Noack et al. review recent applications of the density matrix renormalization group technique to study

Computer Simulation Studies in Condensed-Matter Physics VII
Editors: D.P. Landau · K.K. Mon · H.-B. Schüttler

strongly correlated electron models. Here, they present results obtained with this method for the spin and pairing correlations in coupled Hubbard chain systems. Next, J. K. Freericks and M. Jarrell have used quantum Monte Carlo methods to study electron-phonon systems in infinite dimensions. They present results for the competition between charge density wave and superconducting instabilities and for the crossover between weak- and strong-coupling regimes in the infinite-D Holstein model. A. Muramatsu et al. then discuss their recent quantum Monte Carlo work on the excited state properties of the 1-D Hubbard model. Using maximum entropy methods, they have extracted detailed information about single-particle, spin and charge excitation spectra. The final paper in this section by G. Ortiz reviews recent applications of the fixed-phase method for simulating 2D fermion systems in strong magnetic fields. He presents new results for the stability of the quantum Hall liquid and for the transition to the 2D electron Wigner crystal in a magnetic field.

The third section of the proceedings contains a single manuscript. In a more nontraditional contribution for this Workshop, R. Gerling deals with the increasingly important problem of computer security. Encryption algorithms are described along with their relative strengths and weaknesses, and the importance of random number generators is made evident.

In the first contributed paper, M. A. Novotny introduces a new Monte Carlo algorithm (MCAMC) which is particularly useful for slowly relaxing systems. R. B. Pandey discusses non-Arrhenius conductivity in a driven system of interacting lattice gas. Monte Carlo simulation is used to evaluate the conductivity which shows a nonmontonic dependence on the concentration. K. F. Bassler then describes several non-equilibrium steady state Ising models. The resultant phase diagrams and critical behavior in these unusual systems are discussed. B. Dünweg introduces several Fourier accelerated Langevin simulation algorithms which are useful for the study of two-dimensional ring polymers. Both speed and accuracy are discussed. In an application to bioplymers, U. H. E. Hansmann & Y. Okamoto show how multicanonical simulations can be applied to non-polar amino acids. They observe direct folding of helices from random initial states. F. Schneider et al. have studied 2-D classical and quantum fluids using the Gibbs ensemble Monte Carlo methods. They present results for the phase separation and gas-liquid transitions in classical hard disc fluids and fluids with internal quantum degrees of freedom. J. Lee and M. Y. Choi present a paper on optimization by multicanonical annealng and the traveling salesman problem. The method combines multicanonical technique and an annealing scheme. Results for the traveling salesman problem are given. B. M. Gorman and C. C. A. Günther apply field-theoretical and transfer-matrix methods to study homogeneous nucleation in Ising ferromagnets. Free-energy density computed from transfer matrix method is compared to field-theoretical and Monte Carlo results. G. S. Canright and G. I. Watson describe a new method for finding complex metastable states in frustrated classical systems. Based on a dissipative dynamical map, the method gives energy states which are competitive with the groundstate. P. Lallemand considers representation of relaxation phenomena by lattice gas models and discusses two-dimensional fluid flow simulation with frequency dependent shear or bulk viscosity. G. A. Baker and J. J. Erpenbeck show how standard Monte Carlo simulations of the 2d can be speeded up using the Markov property method. They also estimate the renormalized coupling constant for the 3d Ising model. The paper by C. Sisson discusses the decoupled cell Monte Carlo method and its application in calculating groundstate energies of quantum systems. A specific implementation for the 1-D XY-model is described. G. M. Buendia et al. report on a computer study of a two-dimensional mixed Ising ferromagnet. Monte Carlo simulations, ground-state analysis, and numerical transfer-matrix method are used. The results indicate an Ising-like transition with no compensation point and no tricritical point. In the final paper of this section, L. Pratt and I. M. Khan report on computational studies of organolithium compounds. They present a comparison of results for the bond lengths and dimerization energies obtained by *ab initio*, density functional and semi-empirical methods.

Part I

Classical Systems

Complexity in Rayleigh-Bénard Convection

H.W. Xi[1], *J.D. Gunton*[2], *and J. Viñals*[3]

[1]Exxon Research and Engineering Company, Annandale, NJ 08801, USA
[2]Physics Department, Lehigh University, Bethlehem, PA 18015, USA
[3]Supercomputer Computations Research Institute, B-186,
Florida State University, Tallahassee, FL 32306-4052, USA and
Department of Chemical Engineering, B-203, FAMU/FSU
College of Engineering, Tallahassee, FL 32316-2185, USA

We outline a simple derivation of a generalized Swift-Hohenberg model (GSH) and present a numerical study of mean-flow induced spiral turbulence for this model in two dimensions. We show that this model has a spatiotemporal chaotic state comprised of a large number of rotating spirals. These results are quite similar to recent experimental observations.

1 Introduction

One of the most striking examples of spatiotemporal self-organized phenomena in non-equilibrium systems is a spontanous formation of the spiral state observed in chemical and biological excitable systems [1]. An excitable system is relatively stable when subjected to a small perturbation; large perturbations, on the other hand, drive such a system into large excursions from its initial state. A well-known excitable system is the Belousov-Zhabotinsky (BZ) reaction [2]. In this reaction, spiral waves of oxidation radiate through a chemical mixture, and the waves are in phase along the spiral. The formation of spiral patterns in the BZ system results from the coupling of reaction and a transport process such as diffusion, and has been extensively studied, both theoretically and experimentally during the past decade.Remarkably, similar spatiotemporal spiral patterns have been observed quite recently in convection in Rayleigh-Bénard systems [3, 4, 5]. There is a transition from a global ordered parallel state to a spatiotemporal chaotic state consisting of many localized coherent spiral structures. These remarkable experimental observations of spiral pattern formation provide a challenge for the theorist to explain. We have numerically studied a generalized Swift-Hohenberg model (GSH), described below, and have shown for the first time spatialtemporal spiral pattern formation in Rayleigh-Bénard convection[5]. Our model results are in good qualitative agreement with recent experimental studies [4].
Spatiotemporal chaos is a generalization of temporal chaos, and results when the number of spatial degrees of freedom becomes quite large [6]. This is in contrast to temporal chaos, in which the spatial degrees of freedom are few. As a consequence the strange attractor that presumably characterizes spatialtemporal chaos lies in a large dimensional phase space. The fractal dimension that characterizes this strange attractor has been conjectured to be proportional to $(L/\xi)^d$, where L is the linear size of the system, ξ is a correlation length and d is the dimensionality. It is much more difficult to characterize spatiotemporal chaos in a quantitative manner than for the case of temporal chaos. Conversely, this complexity of convection in fluid systems as manifest in spiral chaos makes this field a fascinating new subject for study.
In Section II, we give a derivation of a generalized Swift-Hohenberg model equation to explain pattern formation in Rayleigh-Bénard fluid systems. In Section III we briefly outline the numerical algorithm for solving the GSH equations. In Section IV we summarize some numerical results obtained for the GSH model. In Section V, we give a brief conclusion.

Springer Proceedings in Physics, Vol. 78
Computer Simulation Studies in Condensed-Matter Physics VII
Editors: D.P. Landau · K.K. Mon · H.-B. Schüttler

2 Mathematical formulation of the problem

Rayleigh-Bénard convection in monocomponent fluids is in general governed by the full three dimensional Navier-Stokes equations. These are often further simplified by using the well-known Boussinesq approximation. Because of the difficulty in solving the initial value problem posed by the fluid equations for large aspect ratio systems, many authors have focused on the study of simpler two dimensional model equations. An example of such a model is the so-called "Swift-Hohenberg" (SH) equation, which is asymptotically equivalent to the long distance and long time behavior of the fluid equations near onset of convection and in the Boussinesq approximation [7]. A great deal of theoretical and numerical work on this latter model has been done by Cross [8], and by Greenside et al. [9]. We give below a simple derivation of a generalized Swift-Hohenberg model that includes an explicit coupling to large scale mean flow, following an approach due to Manneville [10]. This model is the basis of our study of spatiotemporal chaos.

2.1 Boussinesq approximation

We consider a fluid layer of finite thickness and infinite horizontal extent, which is heated from below. For the dimensionless formulation of the problem we introduce the thickness d of the layer as the length scale, d^2/κ as the time scale, κ/d as the velocity scale, $\rho_0\kappa^2/d^2$ as the pressure scale, $\kappa\nu/\alpha g d^3$ as the temperature scale and ν/d^2 as the rotation frequency scale. Here ρ_0 is the fluid density, κ is the thermal diffusivity, ν is the kinematic viscosity and α is the thermal expansion coefficient. The Boussinesq approximation gives the following for the equations of motion and the heat equation,

$$\nabla \cdot \vec{u} = 0, \tag{1}$$

$$\frac{\partial \vec{u}}{\partial t} + (\vec{u} \cdot \nabla)\vec{u} = -\nabla p + Pr\theta \hat{e}_z + Pr\nabla^2 \vec{u}, \tag{2}$$

$$\frac{\partial \theta}{\partial t} + (\vec{u} \cdot \nabla)\theta = RW + \nabla^2 \theta, \tag{3}$$

where $\vec{u} = (U, V, W)$ is the fluid velocity, and the Rayleigh number and Prandtl number are defined as usual, $R=\alpha\Delta T g d^3/\nu\kappa$, $Pr=\nu/\kappa$. θ denotes the deviation from the imposed conduction profile temperature.

By taking the vertical components of the curl and of the curl curl of the Navier-Stokes equation, we obtain

$$\frac{\partial \omega}{\partial t} = Pr\nabla^2 \omega - NL_1 \ , \tag{4}$$

$$\frac{\partial}{\partial t}\nabla^2 W = Pr\nabla_\perp^2 \theta + Pr\nabla^2\nabla^2 W - NL_2 \ , \tag{5}$$

$$\frac{\partial \theta}{\partial t} = RW + \nabla^2\theta - NL_3 \tag{6}$$

where the vertical component of the vorticity is $\omega = (\nabla \times \vec{u})_z$ and the nonlinear terms (NL_1, NL_2 and NL_3) are

$$NL_1 = [\nabla \times (\vec{u} \cdot \nabla \vec{u})]_z \ , \tag{7}$$

$$NL_2 = -[\nabla \times \nabla \times (\vec{u} \cdot \nabla \vec{u})]_z \ , \tag{8}$$

$$NL_3 = (\vec{u} \cdot \nabla)\theta. \tag{9}$$

Operating with $(\partial/\partial t - \nabla^2)$ on Eq. (5) and eliminating θ from the linear parts of Eq. (6), we obtain

$$(\frac{\partial}{\partial t} - \nabla^2)\nabla^2(\frac{\partial}{\partial t} - Pr\nabla^2)W = RPr\nabla_\perp^2 W + NL \ , \tag{10}$$

where

$$NL = -(\frac{\partial}{\partial t} - \nabla^2)NL_2 - Pr\nabla_\perp^2 NL_3 \tag{11}$$

2.2 Generalized Swift-Hohenberg model (GSH)

For simplicity we use free-free boundary conditions of vanishing tangential stress and infinite conductivity at the top (z=1) and the bottom (z=0), which can be expressed in the form

$$\theta = W = \partial_{zz}^2 W = 0 \tag{12}$$

According to the Galerkin expansion method, the solution of the Boussinesq equations can be expressed as a superposition of a set of basis functions

$$(U, V, \omega) = \sum_n [U_n(x,y,t), V_n(x,y,t), \omega_n(x,y,t)] \cos(n\pi z), \tag{13}$$

$$(W, \theta) = \sum_n [W_n(x,y,t), \theta_n(x,y,t)] \sin(n\pi z). \tag{14}$$

The critical mode corresponds to $n = 1$ and the critical Rayleigh number and the critical wavevector can be shown to be $R_c = 27\pi^4$ and $q_c = \pi/\sqrt{2}$ respectively [6]. To the lowest order (n=1) near onset, we have

$$U_1 = a\frac{\partial W_1}{\partial x}, \; V_1 = a\frac{\partial W_1}{\partial y}, \; \theta_1 = cW_1 \quad , \tag{15}$$

where

$$a = \frac{\pi}{q_c^2}, \; c = \frac{(q_c^2 + \pi^2)^2}{q_c^2} \tag{16}$$

We now introduce a small parameter $\epsilon = (R - R_c)/R_c$ and use ϵ for a perturbation expansion. We define the operator [10] $\mathcal{L} = \partial_{xx}^2 + \partial_{yy}^2 + q_c^2$, which is of order $\epsilon^{\frac{1}{2}}$, and ∂_t, which is of order ϵ. We take advantage of the fact that $\partial_{zz}^2 = -\pi^2 = -2q_c^2$, and set $\nabla^2 = \mathcal{L} - 3q_c^2$. The essential approximation involved in the derivation of the amplitude equations is a truncation of the expansion to retain only the lowest mode (n=1) and to limit the horizontal wavevectors $\vec{q}$ to the vicinity of the circle $|\vec{q}| = q_c$. Using this approximation, we expand Eq.(10) using ϵ as a small parameter. The lowest order in this expansion yields the equation which governs the horizontal dependence of W_1 near onset,

$$\tau_0 \frac{\partial W_1}{\partial t} = \epsilon W_1 - \frac{\xi_0^2}{4q_c^2}\left(\nabla_\perp^2 + q_c^2\right)^2 W_1 - NLW, \tag{17}$$

where

$$\xi_0^2 = \frac{4}{3q_c^2}, \tag{18}$$

$$\tau_0 = \frac{Pr + 1}{3Prq_c^2}, \tag{19}$$

$$NLW = \left[(-\nabla^2 NL_2 + Pr\nabla_\perp^2 NL_3)/(3PrR_c q_c^2)\right]|_{n=1}. \tag{20}$$

The notation $[...]|_{n=1}$ means that one projects the function in brackets onto $\sin(\pi z)$. In the limit of large Prandtl number $Pr \gg 1$, Eq. (20) takes the form

$$NLW = \left[(\vec{U}_0 \cdot \nabla_\perp)\theta_1 + (1/2)\left[U_1(\partial\theta_2/\partial x) + V_1(\partial\theta_2/\partial y) - 2\pi W_1\theta_2\right]\right]/(3R_c q_c^2) \tag{21}$$

where

$$\vec{U}_0 = (\partial\xi/\partial y)\hat{e}_x - (\partial\xi/\partial x)\hat{e}_y, \quad and \;\; \omega_0 = -\nabla_\perp^2\xi, \tag{22}$$

where ξ is the mean flow field. Next we need to know the equation for the second order harmonics modes(n=2) of θ_2 and the zero order modes of ω_0. First, we assume that the second harmonics (n=2) are expected to follow the critical mode (n=1) in an adiabatic way(i.e. $\partial W_2/\partial t = \partial\theta_2/\partial t = 0$). With this simplification the equations for W_2 and θ_2 following from Eq. (5) and Eq. (6) are

$$\nabla_\perp^2\theta_2 + (\nabla_\perp^2 - 4\pi^2)^2 W_2 = NL_2|_{n=2} = (1/Pr)\left[\frac{1}{2}(a+\pi a^2)\nabla_\perp^2\mathcal{M}_1\right] \quad , \tag{23}$$

$$R_c W_2 + (\nabla_\perp^2 - 4\pi^2)\theta_2 = NL_3|_{n=2} = \frac{1}{2}ac\mathcal{M}_1 \quad , \tag{24}$$

where

$$\mathcal{M}_1 = q_c^2 W_1^2 + (\nabla_\perp W_1)^2 \tag{25}$$

Here we shall take advantage of the fact that for a single set of straight rolls, $\mathcal{M}_1$ is equal to q_c^2 times the squared modulus of the amplitude of W_1, so that we can replace $\nabla_\perp^2$ by 0 everywhere in Eq. (23) and Eq. (24). We thus obtain:

$$W_2 \approx 0, \tag{26}$$

$$\theta_2 \approx (-1/4\pi^2)\frac{1}{2}ac\mathcal{M}_1. \tag{27}$$

Substituting Eq. (27) into Eq. (17) and using Eq. (15), we get

$$\tau_0\partial_t W_1 + (c/3p_c^4)(\vec{U}_0\cdot\nabla)W_1 = \epsilon W_1 - (\xi_0^2/4q_c^2)\left(\nabla_\perp^2 + q_c^2\right)^2 W_1 - g_3 W_1(1/q_c^2)\mathcal{M}_1 \quad , \tag{28}$$

We note that if one negelect the term $(\vec{U}_0\cdot\nabla)W_1$ in Eq. (28), then this equation is the equation of motion first derived by Swift and Hohenberg [7].
We next note a crucial point due to Siggia and Zippelius [11] who point out that the term n=0 in the Galerkin expansion Eq.(13),(i.e. the term which is independent of z) should not be omitted in the general case. The dynamics of the horizontal large-scale flow $\vec{U}_0(x,y,t) = (U_0, V_0)$ calculated from Eq. (4) thus can be written as

$$(\frac{\partial}{\partial t} - Pr\nabla_\perp^2)\nabla_\perp^2\xi = NL_1|_{n=0} = (a^2/2)\mathcal{M}_2 \quad , \tag{29}$$

where

$$\mathcal{M}_2 = (\nabla_\perp\nabla_\perp^2 W_1 \times \nabla_\perp W_1)\cdot\hat{e}_z. \tag{30}$$

Finally, dropping the index 1 from W_1, and 0 from $\vec{U}_0$ and gathering all the terms, we obtain

$$\tau_0\partial_t W + (c/3p_c^4)(\vec{U})\cdot\nabla)W = \epsilon W - (\xi_0^2/4q_c^2)\left(\nabla_\perp^2 + q_c^2\right)^2 W - g_3 W(1/q_c^2)\mathcal{M}_1 \quad , \tag{31}$$

$$(\frac{\partial}{\partial t} - Pr\nabla_\perp^2)\nabla_\perp^2\xi = (a^2/2)\mathcal{M}_2 \quad , \tag{32}$$

where $g_3 = (c/24p_c^4)$. To simplify the notation, we now rescale Eq. (31) and Eq. (32) by letting $\vec{r}'=q_c\vec{r}$, $t' = (q_c^2\xi_0^2/4\tau_0)t$, $\epsilon' = (4/q_c^2\xi_0^2)\epsilon$, $\psi = (2\sqrt{g_3}/q_c\xi_0)W$ and $\xi' = (2g_3q_c^2/\tau_0\pi^2)\xi$. After dropping the index prime ($'$), we obtain

$$\partial_t\psi + g_m(\vec{U}\cdot\nabla)\psi = \epsilon\psi - \left(\nabla_\perp^2 + 1\right)^2\psi - \psi\mathcal{M}_1 \quad , \tag{33}$$

$$(\partial_t - Pr^*\nabla_\perp^2)\nabla_\perp^2\xi = \mathcal{M}_2 \tag{34}$$

where $g_m = (16\tau_0\pi^2/q_c^2\xi_0^2)$, and $Pr^* = (4\tau_0/\xi_0^2)Pr$

$$\mathcal{M}_1 = \psi^2 + (\nabla_\perp\psi)^2 \tag{35}$$

$$\mathcal{M}_2 = (\nabla_\perp\nabla_\perp^2\psi \times \nabla_\perp\psi)\cdot\hat{e}_z. \tag{36}$$

Equations (33)-(36) define a generalized Swift-Hohenberg model in which an important coupling of the order parameter ψ to the horizontal velocity field $\vec{U}$ (the so-called mean drift velocity) is included. We will see later that this coupling is essential for the formation of the spiral chaos state mentioned in the introduction.

2.3 Convection in a layer with rigid-rigid boundaries

Our analysis until now has been restricted to the free-free boundary conditions. However, the experimental case correspond to rigid-rigid boundary conditions. Since rigid top and bottom boundary conditions do not permit a z-independent uniform horizontal flow, the mean drift generation by a modulation of the roll pattern is expected to be of higher order in ϵ. The complete calculation for rigid boundaries conditions has never been carried out due to its complexity. We will therefore follow an argument by Siggia and Zippelius [11], and limit ourselves to a phenomenological model. Let us assume that the vertically integrated horizontal drift velocity

$$\vec{\mathcal{V}}(x,y,t) = \int_0^1 dz\vec{U}(x,y,z,t) \tag{37}$$

enters the equation for $\vec{U}$ in Eq. (33) as before. Because of incompressibility, $\vec{\mathcal{V}}(x,y,t)$ is divergence-free and gives rise to a mean flow field $\xi(x,y,t)$. The simplest way to correct Eq. (34) is to note that between rigid boundaries, the slowly varying horizontal drift velocity $\vec{U}(x,y,z,t)$ must vanish on the two surfaces. When the diffusive Laplacian in Eq. (4) acts on the vorticity ω, there will be a term arising from the derivatives with respect to z. The evalution of the right-hand- side Eq. (34) should remain essentially unchanged except for some complicated functions of z, which we can eliminate by integration over z. The final results of a generalized Swift-Hohenberg equation can be written as:

$$\partial_t\psi + g_m(\vec{\mathcal{V}}\cdot\nabla)\psi = \epsilon\psi - \left(\nabla_\perp^2 + 1\right)^2\psi - \psi^3 \tag{38}$$

$$\left[\partial_t - Pr^*(\nabla_\perp^2 - c^2)\right]\nabla_\perp^2\xi = (\nabla_\perp\nabla_\perp^2\psi \times \nabla_\perp\psi)\cdot\hat{e}_z. \tag{39}$$

where c^2 is an unknown constant resulting from the integration of $\partial^2\omega/\partial^2 z$ over z and where the drift velocity is

$$\vec{\mathcal{V}} = (\partial\xi/\partial y)\hat{e}_x - (\partial\xi/\partial x)\hat{e}_y. \tag{40}$$

Equations (38-40) define the generalized Swift-Hohenberg model (GSH) which we study below. We note that as in Eqs. (33) and (34) we have dropped the prime index in Eqs. (38) and (39). We also note that the cubic term in Eq. (38) is simpler than that in Eq. (33), due to the approximation. Finally, since this is a phenomenological model, we use it for small Prandtl number, which is an additional assumption.

3 Numerical method for the model equation

We now briefly summarize the numerical algorithm [13] which we use to solve the GSH equations Eq. (38)-(40). The key step is to recognize that the GSH equation can be solved by the repeated solution of the *linear* constant-coefficient biharmonic equation,

$$(\nabla^2\nabla^2 + a\nabla^2 + b)u = f_1, \tag{41}$$

with boundary conditions,

$$u|_R = f_2, \quad \hat{n}\cdot\nabla u|_R = f_3, \tag{42}$$

on a circular domain R. The constants a and b are real numbers, $\hat{n}\cdot\nabla$ denotes the normal derivative taken at the boundary of the domain, and f_1 , f_2 and f_3 are given functions. For the case of rigid boundaries, we have $f_2 = f_3 = 0$. Consider an *implicit* backward Euler discretization scheme in time, to yield the following finite difference set of equations,

$$\frac{\psi(\tau+\Delta\tau)-\psi(\tau)}{\Delta\tau} + g_m\vec{U}(\tau+\Delta\tau)\cdot\nabla\psi(\tau+\Delta\tau) =$$
$$L\psi(\tau+\Delta\tau) - \psi^3(\tau+\Delta\tau), \tag{43}$$

$$\frac{\nabla^2\left[\xi(\tau+\Delta\tau)-\xi(\tau)\right]}{\Delta\tau} + Pr\nabla^2\left[c^2-\nabla^2\right]\xi(\tau+\Delta\tau) =$$
$$-\left[\nabla\psi(\tau+\Delta\tau)\times\nabla(\nabla^2\psi(\tau+\Delta\tau))\right]\cdot\hat{e}_z, \tag{44}$$

where $\psi(\tau)$ and $\xi(\tau)$ are the known solutions of Eqs. (38)-(39) at time τ, $\Delta\tau$ is the time step, $L = \epsilon-(1+\nabla^2)^2$ is a linear biharmonic operator, and $\psi(\tau+\Delta\tau)$ and $\xi(\tau+\Delta\tau)$ are the unknown implicit solutions at the next time step $(\tau+\Delta\tau)$ (we have temporarily suppressed the spatial indices). We solve Eqs. (43) and (44) by using a multi-iteration Gauss-Seidel scheme. We assume that $\psi(\tau+\Delta\tau)$ and $\xi(\tau+\Delta\tau)$ are obtained by successive approximations of the form,

$$\psi(\tau+\Delta\tau) \simeq \psi_{k+1} = \psi_k + \delta_k, \qquad \psi_0 = \psi(\tau), \tag{45}$$
$$\xi(\tau+\Delta\tau) \simeq \xi_{k+1} = \xi_k + \theta_k, \qquad \xi_0 = \xi(\tau), \tag{46}$$

where $\psi_k(x,y)$ and $\xi_k(x,y)$ are approximations at the $k-th$ iteration. The outer correction fields δ_k and θ_k are assumed to be small ($\|\delta_k\| \ll \|\psi_k\|$ and ($\|\theta_k\| \ll \|\xi_k\|$) in the maximum norm. By substituting Eqs. (45) and (46) into Eqs. (43) and (44) and linearizing in the small fields δ_k and θ_k, we obtain the standard Gauss-Seidel iteration scheme for the unknown corrections δ_k and θ_k,

$$\left[L-3\psi_k^2-\frac{1}{\Delta\tau}\right]\delta_k = \frac{\psi_k-\psi_0(\tau)}{\Delta\tau} + g_m\vec{U}_k\cdot\nabla\psi_k - (L\psi_k-\psi_k^3-g_2\psi_k^2), \tag{47}$$

$$\nabla^2\left[\nabla^2-d^2\right]\theta_k = -\frac{1}{Pr\Delta\tau}\nabla^2\xi_0 - \nabla^2\left[\nabla^2-d^2\right]\xi_k$$
$$+\frac{1}{Pr}\left[\nabla\psi_{k+1}\times\nabla(\nabla^2\psi_{k+1})\right]\cdot\hat{e}_z, \tag{48}$$

with $d^2 = (c^2+\frac{1}{Pr\Delta\tau})$. The right hand side of Eqs. (47) and (48) are the $k-th$ outer residuals, $r^{\psi}_{outer}(k)$ and $r^{\xi}_{outer}(k)$ which measure the extent to which Eqs. (43) and (44) are satisfied by the $k-th$ order approximation. Given the residuals, we solve for the outer corrections, δ_k and θ_k, and thus obtain a better approximation to $\psi(\tau+\Delta\tau)$ and $\xi(\tau+\Delta\tau)$. Iteration continues over the index k, until both the outer residuals and the outer corrections are small compared to ψ_k and ξ_k, that is,

$$max(\|\delta_k\|, \|r^{\psi}_{outer}(k)\|) \le \epsilon_{rel}(\|\psi(k)\|) + \epsilon_{abs}, \tag{49}$$

and,

$$max(\|\theta_k\|, \|r^{\xi}_{outer}(k)\|) \le \epsilon_{rel}(\|\xi(k)\|) + \epsilon_{abs}, \tag{50}$$

where ϵ_{rel} and ϵ_{abs} are the relative and absolute error tolerances, chosen to be 0.1 and 10^{-4} respectively in our calculations. When the convergence criteria are satisfied, ψ_{k+1} and ξ_{k+1} are set to be $\psi(\tau+\Delta\tau)$ and $\xi(\tau+\Delta\tau)$ respectively. To discretize the spatial derivatives for the geometry of interest, we use the usual 5- and 13- point discretizations of the Laplacian and biharmonic operators in finite differences on a spatial grid of $N \times N$ points. This is of second-order accuracy in the mesh spacing.

4 Numerical results

In this section, we present our numerical studies of the GSH model of Eqs(38-40) in a large aspect ratio cell, and compare our results with the spiral pattern evolution obtained in the experiments of Morris et al [4]. We consider as the initial condition ($\psi(\vec{r}, t=0)$) a Gaussian random variable with zero mean and variance 10^{-4}. We numerically solve Eqs (38-40) in a circular cell of radius $r = 32\pi$, which corresponds to an aspect ratio $\Gamma = r/\pi = 32$. A grid with N^2 nodes has been used with spacing Δx=Δy=$64\pi/N$, and $N = 512$. We approximate the circular boundary conditions on ψ and ξ by taking $\psi(\vec{r}, t) = \xi(\vec{r}, t) = 0$ for $||\vec{r}|| \geq D/2$, where $\vec{r}$ is the location of a node with respect to the center of the domain, and D is the diameter of the circular domain. The parameters in the equation have been chosen to approximately match the experiments, i.e. $g_m = 50$ and $c^2 = 2$. We likewise choose the Prandtl number to be equal to 1, which is the value used in the experiment.
As we see from Figure 1, the convective pattern is dominated by global parallel rolls for $\epsilon =$ 0.1, very near onset. This is in agreement with the classical theory of Busse [12]. However, as one increases ϵ, the situation changes dramatically in that parallel rolls no longer form the stationary state. Rather, as shown in Figure 2, for $\epsilon = 0.7$, the system exhibits a time-dependent, spatiotemporal chaotic spiral pattern. This is a novel result, which disagrees with classical theory (Busse [12]), but is remarkably similar in visual appearance to that observed experimentally [4]. The time evolution of this spiral state shows that spirals nucleate, interact and annihilate, yielding a pattern that is disordered on a length scale greater than the average spiral diameter. This can be seen in another snapshot shown in Figure 3.

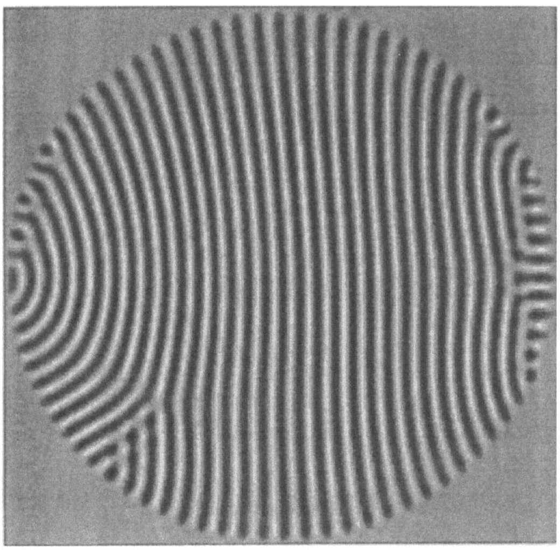

Figure 1. A parallel roll pattern obtained from a random initial condition in a cylindrical cell of aspect ratio $\Gamma = 32$. The values of the parameters used are $\epsilon = 0.1$, $g_m = 50$ and $c^2 = 2$. The field ψ is shown. Dark regions correspond to $\psi > 0$ and light regions to $\psi < 0$.

Figure 2. Typical configuration of a spiral chaos pattern. The configuration shown here has evolved from random initial conditions in a cylindrical cell with aspect ratio $\Gamma = 32$. The values of the parameters used are $g_m = 50$ and $c^2 = 2$. The parameter ϵ is quenched from 0 to 0.7 in the simulation. The configuration shown is at $t = 900 \approx \tau_h$.

A quantitative characterization of this spatiotemporal chaos has been given elsewhere [5] [14]. We summarize here certain key features of our analysis. First, we show in Figure 4 the two dimensional power spectrum $< P(\vec{k}) > = < |\psi(\vec{k},t)|^2 >$ where $\psi(\vec{k},t)$ is the spatial Fourier transform of $\psi(\vec{r},t)$ and $<>$ denotes a time average over two horizontal-diffusion times ($\tau_{horizontal} = \Gamma^2$). The most interesting feature is that the intensity of the broad spectrum appears to be isotropic. The inverse transform of $< P(\vec{k}) >$ is the spatial correlation function, $< C(\vec{r}) >$. The circularly averaged correlation function $C(r)$, shown in Figure 5, exhibits an oscillatory behavior with an envelope function that can be fit by an exponential function. If one uses the exponential function to define a correlation length, one finds $\xi = O(\lambda_c)$, where λ_c is the critical wavelength near onset. Finally, we note that the correlation function which best describes the localized coherent spiral structure is the vorticity-vorticity correlation function. Although we do not show this function here, its behavior reveals a very compact core structure for the spiral.

The most interesting qualitative finding of our numerical work and the recent experiment work [4] is that this interesting spatiotemporal chaotic state is not predicted by classical theory. Thus a challenging open question is a theoretical explanation of this phenomena [15]. We should also note that if we set the coupling to the mean flow field equal to zero (i.e. g_m=0), we do not observe spiral chaos.

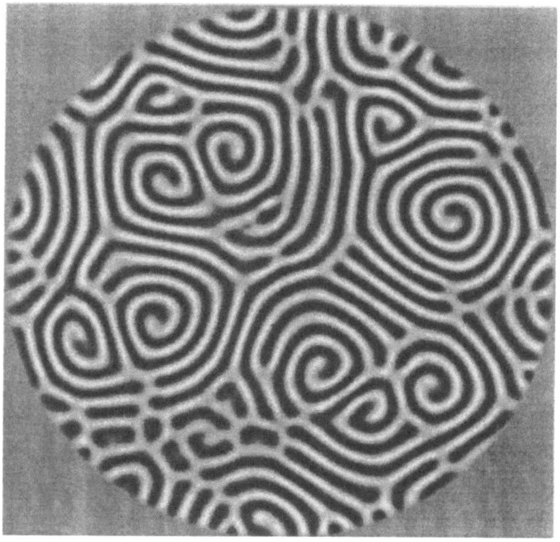

Figure 3. The time evolution of the spiral state at a later time than shown in Figure 2. We see that spirals nucleate, interact and annihilate, yielding a a quite different spiral pattern that is disordered on a length scale greater than the average spiral diameter. The configuration shown is at $t \approx 1700$.

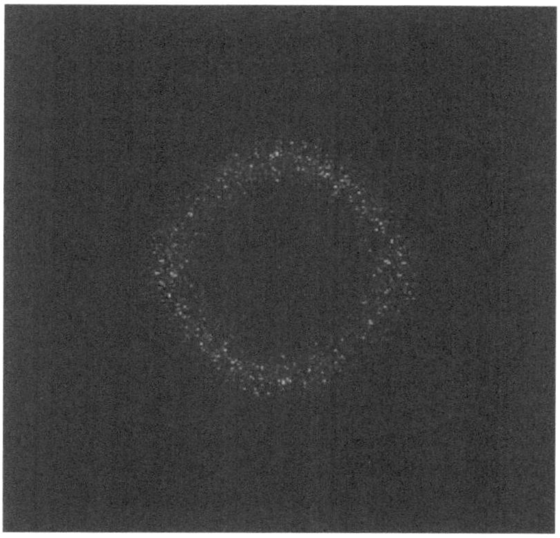

Figure 4. Power spectrum in $\vec{k}$-space of order parameter ψ. It is interesting to note that the intensity of the spectra is approximately isotropic and broad near the critical wavenumber.

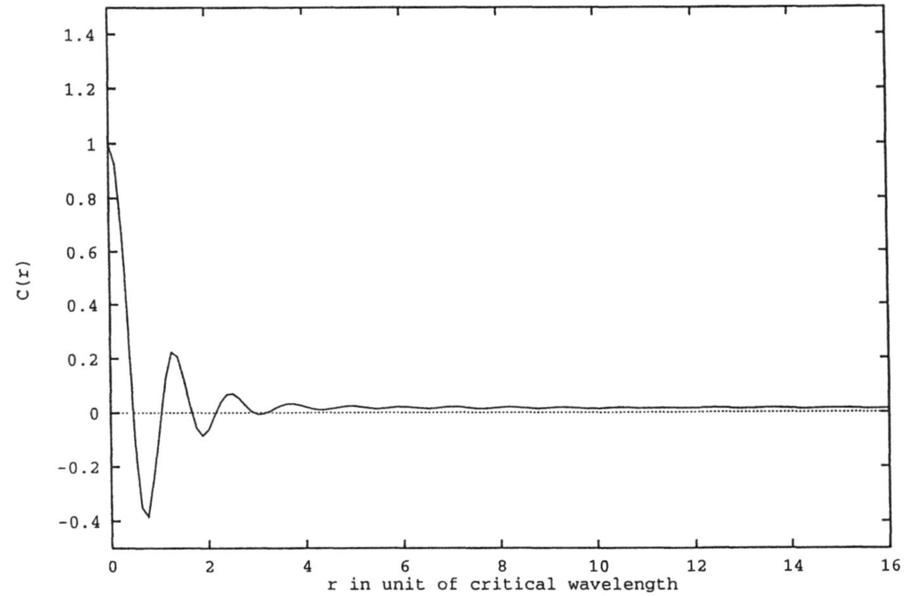

Figure 5. Circularly averaged correlation function $C(r)$. Note that $C(r)$ has an oscillatory behavior with decay of the envelop function. The correlation length $\xi \approx O(\lambda_c)$. Here λ_c is the critical wavelength near onset.

5 Conclusion

We have shown that a generalization of the Swift-Hohenberg model that includes a crucial coupling to large scale mean flows yields a spatiotemporal chaotic state. This spiral state has also been observed in a recent experiment. Both the numerical and experiment results demonstrate the need for a new theory of Rayleigh-Bénard convection (at sufficiently low Prandtl number) that will explain the complex behavior associated with spatiotemporal chaos.

Acknowledgments

This work was supported in part by the National Science Foundation under Grant No. DMR-9100245. This work is also supported in part by the Supercomputer Computations Research Institute, which is partially funded by the U.S. Department of Energy contract No. DE-FC05-85ER25000. The calculations reported here have been carried out on the Cray C90 at the Pittsburgh Supercomputing Center.

References

[1] H.L. Swinney and V.I. Krinsky, ***Waves and Patterns in Chemical and Biological Media***, Physica D **49**, pp.1 1991.

[2] A.T. Winfree, Science **175**, 634 (1972); S.C. Muller, T. Plesser and B. Hess, Science **230**, 661 (1985); W.Y. Tam, W. Horsthemke, Z. Noszticzius and H. Swinney, J. Chem. Phys. **88**,3395 (1988); G.S. Skinner and H. Swinney, Physica D **48**, 1 (1991).

[3] E. Bodenschatz, J.R. de Bruyn, G. Ahlers and D.S. Cannell, Phys. Rev. Lett. **67**, 3078 (1991).

[4] S. W. Morris, E. Bodenschatz, D.S. Cannell and G. Ahlers Phys. Rev. Lett. **71**, 2026 (1993).

[5] H.W. Xi, J.D. Gunton and J. Viñals, Phys. Rev. Lett **71**,2030-2033 (1993).

[6] M.C. Cross and P.C. Hohenberg, Rev.Mod.Phys, **65**, 3 (1993).

[7] J. Swift and P.C. Hohenberg, Phys. Rev. A **15**, 319 (1977).

[8] M.C. Cross, Phys. Fluids **23**, 1727 (1980); Phys. Rev. A **25**, 1065 (1982); Phys. Rev. A **27**, 490 (1983).

[9] H.S. Greenside and W.M. Coughran Jr., and N. L. Schryer, Phys. Rev. Lett. **49**, 726 (1982); H.S. Greenside and W.M. Coughran Jr., Phys. Rev. A **30**, 398 (1984); H.S. Greenside and M.C. Cross, Phys. Rev. A **31**, 2492 (1985).

[10] P.Manneville, J.Physique **44**, 759(1983).

[11] E.D. Siggia and A. Zippelius, Phys. Rev. Lett. **47**, 835 (1981); A. Zippelius and E.D. Siggia, Phys. Rev, A **26**, 178 (1982).

[12] F.H. Busse, J. Fluid Mech. **30**, 625 (1967).

[13] P. E. Bjϕrstad, W. M. Coughran, Jr, H. S. Greenside, D. J. Rose and N. L. Schryer *Elliptic Problem Solvers II* , (Academic Press, New York, 1984).

[14] H.W. Xi, J.D. Gunton and J. Viñals, manuscript in preparation.

[15] M. Assenheimer and V. Steinberg, Nature, **367**,345, (1994).

Numerical Zero-Temperature Results for the 3*d* Edwards-Anderson Ising Spin Glass

B.A. Berg

Department of Physics, The Florida State University, Tallahassee, FL 32306, USA and
Supercomputer Computations Research Institute (SCRI), Tallahassee, FL 32306, USA

A study of the zero temperature properties of the 3d Edwards–Anderson Ising spin glass, by means of multicanonical simulations is reported. Finite size scaling fits of the data are carried out for two hypothetical scenarios: Parisi mean field theory versus a droplet scaling ansatz. With a zero temperature scaling exponent $y = 0.72 \pm 0.12$ the data are well described by the droplet scaling ansatz. Alternatively, a description in terms of the Parisi mean field behavior is still possible. The two scenarios give significantly different predictions on lattices of size $\geq 12^3$.

1 Introduction

The theoretical problem to determine the equilibrium groundstate structure [1, 2] of spin glasses has remained an important, but elusive, question. There is a wide range of phenomena which have been related to this problem, reaching beyond the magnetic alloys spin glass model were invented for to studies of questions in biology such as the theory of memory and evolution [3]. It is generally agreed that the statistical mechanics of the infinite ranged Sherrington–Kirkpatrick Ising spin glass is essentially understood. The replica–symmetry breaking mean field (MF) scheme discovered by Parisi [1] exhibits infinitely many extremal Gibbs states at very low temperature, forming a space with ultrametric topology, which may be arranged in a generation tree. As the temperature is raised, states within increasing distance from each other coalesce until above a certain freezing temperature T_c the equilibrium state is unique.

More realistic models rely on short range interactions. The simplest prototype of this sort is the Edwards–Anderson Ising spin glass (EAI), defined by the Hamiltonian

$$H_J = - \sum_{<ij>} J_{ij} s_i s_j . \tag{1}$$

Here the sum $< ij >$ goes over nearest neighbors. We consider 3*d* systems with periodic boundary conditions and $N = L^3$ spins $s_i = \pm 1$. The exchange interactions between the spins are taken $J_{ij} = \pm 1$, randomly distributed over the lattice such that the constraint $\sum_{<ij>} J_{ij} = 0$ is fulfilled. The subscript J denotes the realization $\{J_{ij}\}$. Extensive numerical studies of this model were carried out in the past [4, 5, 6, 7]. In essence, these simulations seem to have established the existence of a freezing phase transition, although not beyond all doubts [8].

After the Sherrington-Kirkpatrick model had been solved, it became a question of central theoretical interest whether, well below the freezing temperature $T_c = \beta_c^{-1} \approx 1.2$ [6, 7], the 3*d* EAI spin glass model exhibits Parisi MF behavior. It was answered in the negative by proponents of a simple scaling ansatz [9, 10, 11, 12, 13, 14]. These droplet scaling (DS) theories suggest that no more than two pure states (related via a global flip)

Springer Proceedings in Physics, Vol. 78
Computer Simulation Studies in Condensed-Matter Physics VII
Editors: D.P. Landau · K.K. Mon · H.-B. Schüttler

exist at any temperature. The MF approximation is surely valid for $d \to \infty$. It has been suggested [14] to hold down to $d = 4$, and that the DS scenario then applies to $d = 3$.

Simulations [15] in a magnetic field H seemed to favor the MF picture rather than the alternative droplet model. However, it has been pointed out that equilibrium at sufficiently low temperatures has not been reached [16], and a recent Monte Carlo (MC) study [17] also disagrees with the conclusions of [15]. A study of contours of the spin glass susceptibility in the (H, T)–plane as means of distinguishing the two pictures was performed in ref. [18], but remained ambiguous in $d = 3$. Finally, a recent investigation of the face-centered cubic lattice [19] is consistent with the existence of an Almeida–Thouless line in the (H, T)–plane and thus favors again the MF scenario.

Altogether the theoretical understanding remained quite unsatisfactory. This has not only been due to the fact that we do not know of adequate analytical methods to cope with the problem, but even numerical studies have met with formidable difficulties: The dynamical stochastic process set up to simulate equilibrium states fails to reach equilibrium in available computer time. Computer studies, even when performed on optimized special purpose computers [5, 6], were confined to rather high temperatures. Multicanonical simulations [20, 21, 22] try to circumvent this notorious ergodicity problem of canonical low temperature spin glass simulations through regular excursions into the disordered phase, *while* staying in equilibrium. Thus this numerical technique may shed new light on the spin glass groundstate problem. In this talk, I present results from extensive recent multicanonical simulations of the $3d$ EAI model which were carried out in collaboration with Hansmann and Celik [23]. The next section introduces the observables used to characterize the spin glass phase. A summary of the numerical results is then given in section 3, and section 4 contains a few concluding remarks.

2 The Observables

For each system there are $(3N)!/[(3N/2)!]^2$ realizations of the quenched random variables $J = \{J_{ij}\}$. A quantity of decisive importance is the probability density $P(q)$ of the Parisi order parameter q:

$$P(q) = [P_J(q)]_{av} = \frac{[(3N/2)!]^2}{(3N)!} \sum_J P_J(q). \tag{2}$$

By $[\cdot]_{av}$ we denote averaging over the realizations J, whereas $\langle\cdot\rangle_T$ denotes the thermal (time) average for a given realization of bonds. For such a fixed realization $P_J(q)$ is the probability density of the overlap

$$q = q_J = \frac{1}{N} \sum_i s_i^1 s_i^2 \bigg|_J . \tag{3}$$

Here s_i^1 and s_i^2 denote two replica (*i.e.* statistically independent configurations) of the realization J at temperature T. Our normalization is $\int_{-1}^{+1} P(q)dq = \int_{-1}^{+1} P_J(q)dq = 1$. Due to the magnetic field being zero we have the symmetry $P(-q) = P(q)$, such that $\int_{-1}^{+1} q^n P(q)dq = 0$ for n odd. We are therefore only interested in averages over the range $0 \le q \le 1$:

$$\overline{|q|^n} = [\langle |q_J|^n \rangle_T]_{av} = 2 \int_0^1 q^n P(q)dq. \tag{4}$$

Of course, $\overline{q^n} = [\langle q^n \rangle_T]_{av} = \overline{|q|^n}$ for n even. For a fixed realization J the analog averages are defined by

$$\overline{|q_J|^n} = \langle |q_J|^n \rangle_T = 2 \int_0^1 q^n P_J(q)dq. \tag{5}$$

In the infinite volume limit the over-all scenario is that for MF theory $P(q)$ takes nonzero values in a continuous range $|q| < q_1$ with peaks at $\pm q_{\max}$ and $0 < q_{\max} \leq q_1$. We denote the value at the peak by $P_{\max} = P(q_{\max}) = P(-q_{\max})$. If the DS picture is correct, only a double peak structure survives in the infinite volume limit, *i.e.* $\lim_{L\to\infty} P(q) = 0$ for $q \neq q_{\max}$, where $q^{\infty}_{\max}$ is the infinite volume limit of $q^{L}_{\max}$. Whenever without confusion possible, we drop the subscript denoting the lattice size.

Let us now discuss moments of the order parameter distribution. For the spin glass susceptibility $\chi_q = N\overline{q^2}$ the MF as well as the DS scenario suggest divergence $\chi_q \sim N$ for $T < T_c$. However they give significantly different predictions for the variance

$$\sigma^2(\overline{|q|}) = 2\int_0^1 (\overline{|q|} - q)^2 P(q)dq = \overline{q^2} - \overline{|q|}^2. \tag{6}$$

Let us temporarily confine the discussion to zero temperature. In the limit $L \to \infty$ one has $\sigma^2(\overline{q}) \to$ *finite* in MF theory, while $\sigma^2(\overline{q}) \sim L^{-y} \to 0$ within the DS approach. Here $y = -y_T$ is the zero temperature scaling exponent [10, 12], denoted θ in [11, 13]. Within the DS ansatz it is supposed to govern quite generally the finite size scaling (FSS) corrections of expectation values. For instance for the moments $\overline{|q|^n}_L - \overline{|q|^n}_\infty \sim L^{-y}$, or for the position of the maximum $q^L_{\max} - q^{\infty}_{\max} \sim L^{-y}$. For MF theory we assume for most of our fits 1/Volume corrections. The assumption of exponentially small volume corrections may be even more appropriate. However, this introduces one more fit parameter and presently we have not sufficiently many lattice sizes to allow its determination self-consistently. By this reason exponentially small correction are only explored once (for the energy).

Assuming DS, the above zero temperature scaling exponent should also govern the fall-off of the probability density $P(q)$ for q away from $q^{\infty}_{\max}$. For instance $P_L(0) \sim L^{-y}$. However, numerically it is rather tedious to get a sufficiently good statistics at a particular q value. It is more convenient to rely on the probability distribution

$$x(q) = 2\int_0^q P(q')dq', \tag{7}$$

where $q < q_{\max}$ should stay sufficiently far away from $q_{\max}$. It is clear that the FS behavior is still $x_L(q) \sim L^{-y}$, but the statistical noise will be considerably suppressed.

Lack of self-averaging is one prominent feature of MF behavior. This means physical quantities Q exist for which

$$\lim_{L\to\infty} \left[\left(\overline{Q} - \overline{Q}_J \right)^2 \right]_{av} \neq 0 \tag{8}$$

holds. In ref. [15] the probability density itself was studied through $[\int (P(q) - P_J(q))^2 dq]_{av}$. This was criticised by the authors of [16] on the reason that the probability density is so singular that it would be non self-averaging in the DS picture too. Following their suggestion we estimate

$$\sigma^2_J(\overline{q^2}) = \left[(\overline{q^2} - \overline{q^2}_J)^2 \right]_{av} = \left[(\overline{q^2_J})^2 \right]_{av} - \left(\overline{q^2} \right)^2. \tag{9}$$

Again, this quantity stays finite in MF theory, but drops off $\sim L^{-y}$ in the DS picture. In contrast to this, standard thermodynamical quantities like the energy per spin e are self-averaging in both scenarios:

$$e = N^{-1} \left[\langle H_J \rangle_T \right]_{av} = N^{-1} \langle H_J \rangle_T. \tag{10}$$

3 Numerical Results

With the development of multicanonical techniques for disordered systems [22] it has become feasible to generate spin glass groundstates in thermal equilibrium; see [24] for the earlier umbrella sampling, and [25, 26] for the related simulated tempering. A pilot study for the model at hand has been presented in [27]. In essence a multicanonical spin glass simulation proceeds in three steps. First Monte Carlo weights are recursively constructed which will allow to simulate an ensemble, the "multicanonical", which yields canonical expectation values in the temperature range $0 \leq T \leq \infty$ through use of the spectral density. Secondly, equilibrium configurations with respect to the multicanonical ensemble are generated by means of standard MC. In a third step canonical expectation values at desired temperatures are obtained from the analysis.

In this paper we focus on the investigation of groundstate properties. A lower bound on the number of statistically independent groundstates sampled is obtained by counting how often the system moves from the energy $E \geq 0$ region to the groundstate energy $E_{\min}$, and back to the $E \geq 0$ region. This has been termed "tunneling" [21] and we follow this notation, but one should bear in mind that the free energy barriers are actually not overcome by a tunneling process. Our tunneling time τ is the average number of single spin updates corresponding to one tunneling event.

3.1 Statistics and slowing down

We have performed simulations for $L = 4, 6, 8, 12$ ($N = 64, 216, 512, 1728$). For $L \leq 8$ the sum (2) is approximated through 512 randomly chosen realizations of the $\{J_{ij}\}$, whereas we have only 7 realizations for $L = 12$. For all 1,543 cases multicanonical parameters were determined recursively. Then each system was simulated twice with independent random starts and random numbers. This constitutes our two independent replica per realization. In these production runs iterations were stopped when a preset number of tunneling events n_τ had occurred: $n_\tau = 128$ ($L = 4$), 64 ($L = 6$), 32 ($L = 8$) and 10 ($L = 12$). Despite this decrease in tunneling events, the average number of updates per spin n_s (sweeps) did steadily increase. Approximate values are: $n_s = 8 \cdot 10^4$ ($L = 4$), 10^5 ($L = 6$), $7.6 \cdot 10^6$ ($L = 8$) and $50 \cdot 10^6$ ($L = 12$). The average CPU time spent on one $L = 8$ replica was approximately 800 minutes on an IBM 320H workstation.

Per replica we have stored up to 2,048 groundstate configurations. Due to correlations the number of encountered groundstates is, of course, much larger than n_τ. If the number exceeded 2,048, the stored configurations were randomly selected from the total set. For groundstate configuration n this is elegantly done on-line by picking a random integer i_r in the range $1 \leq i_r \leq n$. Configuration n is stored at position i_r if $i_r \leq 2,048$ and discarded otherwise. For both replica the same groundstate energy has to be found between all tunneling counts. This is a strong, albeit not rigorous, criterium to ensure that the correct groundstate energy has not been missed. Indeed, in course of a study [28], aimed at improving the algorithmic performance, lower energy states have been found for a few of the $L = 8$ and $L = 12$ configurations used in this paper. In all $L = 8$ cases the improvement is just by a single energy step, and the indication is that it happens for less than 2% of our recorded configurations. Comparing the groundstate properties of the old and new configurations shows that the presently implied bias favors the MF picture, but is negligible within the statistical noise due to the small number of configurations affected. For $L = 12$ the problem is more serious, but they are anyhow of limited relevance to the subsequent discussion. No improvements were found for any $L = 4$ or $L = 6$ configurations.

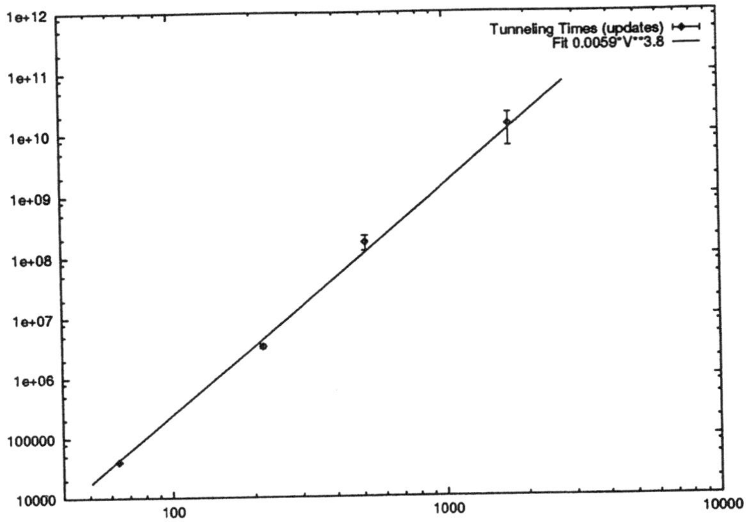

Figure 1: Tunneling time versus lattice size on a double log scale.

Let us first discuss the slowing down as inferred from the FS behavior of the tunneling time τ. A formal, linear fit corresponding to the power law $\tau = a_1 V^{a_2}$ gives $a_2 = 3.8 \pm 0.1$ and is depicted in figure 1. This $\sim V^{3.8}$ slowing down has limited our investigation to rather moderately sized lattices.

3.2 Parisi order parameter distribution

On a semi–log scale figure 2 depicts the obtained zero temperature probability densities (2) for the Parisi order parameter. The $L = 12$ probability density, presented without error bars, is very bumpy due to the small number of realizations, and may only be reliable for a few of the considered physical quantities. Note, altogether the data respect the $P(q) = P(-q)$ symmetry well. For $L = 8$ figure 3 plots the $P_J(q)$ probability densities of two rather extreme $L = 8$ realizations: two peak shape versus continuous distribution. Various different shapes in–between these extremes are also found. From figures 2 and 3 it is evident that only a careful quantitative analysis of these distributions may give hints concerning the $L \to \infty$ groundstate distribution.

Let us now discuss $\sigma^2(\overline{q})$ and $\sigma_J^2(\overline{q^2})$. The DS fit is $a_1 L^{-a_2}$. Fits and data are depicted in figure 4. For each case we give two fit curves. The upper one relies on three data points ($L = 4, 6, 8$), whereas the lower one includes also the $L = 12$ result. When only three data are used, MF and DS fits are both consistent ($Q = 0.10$ and $Q = 0.43$, where Q is the goodness of fit [29]). Once the $L = 12$ data point is included, the consistency of the MF fit becomes marginal ($Q = 0.04$). However, from the $L = 8$ data we have the experience that 10% of the realizations amount to 99% of the $P(0)$ contribution. Consequently, the $L = 12$ data suffer not only from large statistical fluctuations, but are altogether unreliable for quantities which are sensitive to the small q distribution. We therefore rely on the three–point fits for $\sigma^2(\overline{q})$ and $\sigma_J^2(\overline{q^2})$. The two $y = a_2$ estimates are compatible, and we summarize them to

$$y = 0.72 \pm 0.12. \tag{11}$$

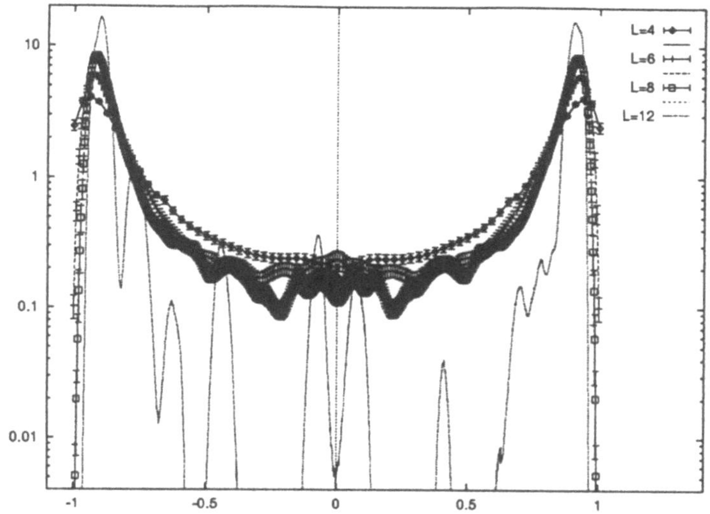

Figure 2: Probability densities $P_L(q)$ for the Parisi order parameter ($L = 4, 6, 8$ and 12).

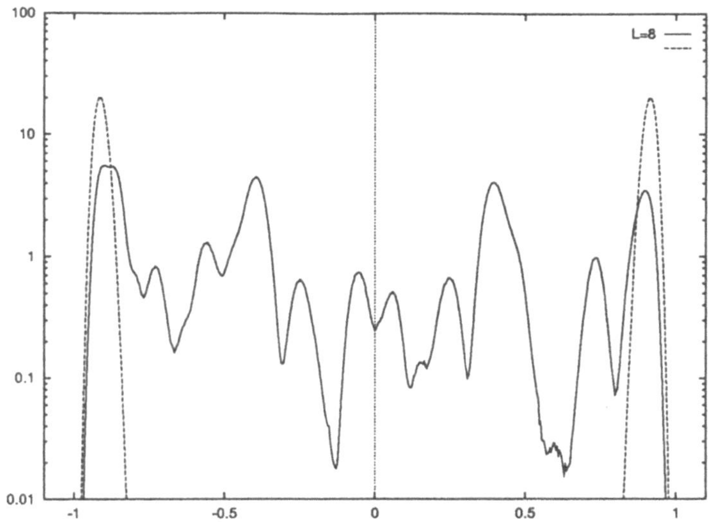

Figure 3: Probability densities $P_J(q)$ for two very different $L = 8$ realizations.

The error bar is not reduced as both estimates rely on the same data set. The above fits were also used for $P(0)$, but the data are too inaccurate to yield meaningful results. As discussed, this is different for the distribution function $x(q)$ with an appropriate choice for q. Our smallest FSS extrapolation of $q_{\max}^{\infty}$ is 0.8791 ± 0.0081. Therefore $q = \frac{1}{2}$ is certainly an appropriate choice. We display the FSS fits for $x(\frac{1}{2})$ in figure 5. Relying on

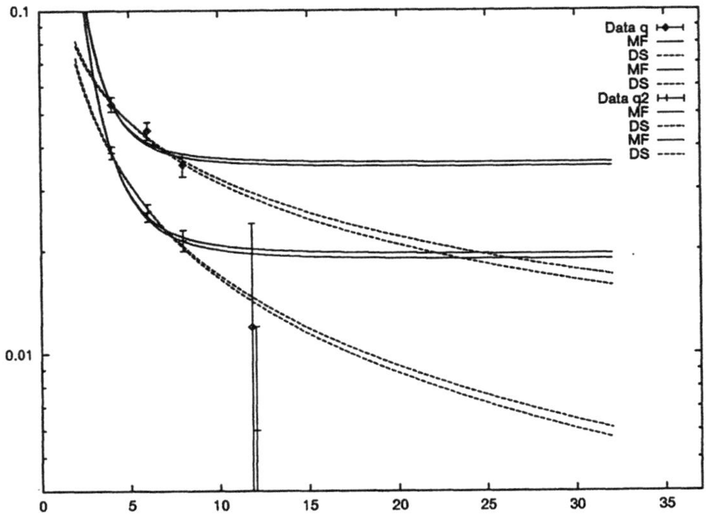

Figure 4: Fits for $\sigma^2(\overline{q})$ (Data q) and $\sigma_J^2(\overline{q^2})$ (Data $q2$).

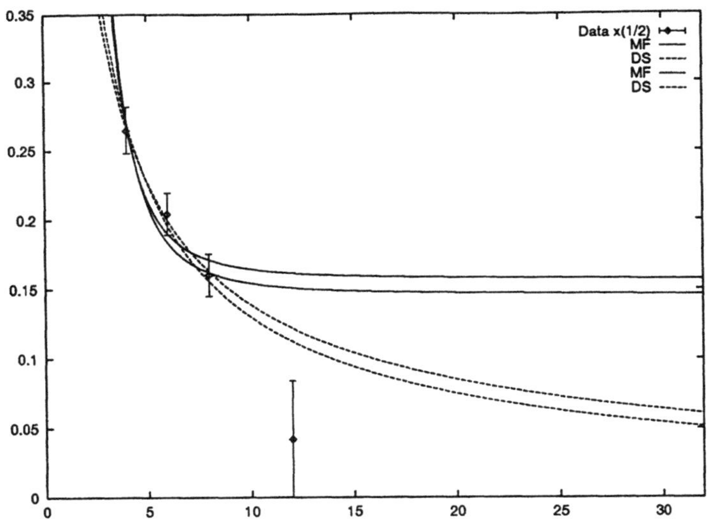

Figure 5: Fits for $x(\frac{1}{2})$.

three lattice sizes again, it is quite satisfying that the obtained zero temperature exponent $y = 0.71 \pm 0.16$ is in excellent agreement with equation (11).

Assuming that our $L = 4-8$ lattices show already typical scaling behavior, we conclude from figures 4 and 5 that similarly accurate data on lattices up to size $L = 16$ would discriminate between the MF and the DS ansatz. Due to the $\sim V^{3.8}$ slowing down the

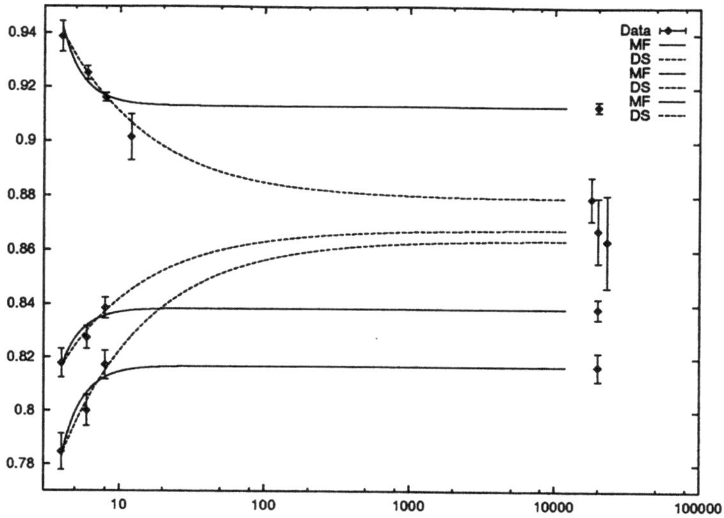

Figure 6: Fits and extrapolations for $\overline{q}_{\max}$ (up), $(\overline{q^2})^{-1/2}$ (middle) and $\overline{q}$ (down).

needed CPU time would be about 3,000 times larger than the one spent on the present investigation. With upcoming massively parallel devices in the teraflop range such a factor can be achieved.

The DS fits for $\overline{q}, \overline{q}^2$ and $q_{\max}$ are of the form $a_1 + a_2 L^{-y}$, where the exponent $y = 0.72$ (11) is now used as input. The corresponding MF fits are of the form $a_1 + a_2 L^{-3}$. A relevant consistency check for the correctness of the DS picture is that the infinite volume estimates of $\overline{q}_{\max}$, $\sqrt{\overline{q^2}}$ and $\overline{q}$ have to agree. Figure 6 shows the MF and DS fits for these quantities. For L a log-scale is used to exhibit $L \to \infty$ clearly, and the infinite volume estimates are depicted towards the end off the scale. With the previously determined zero temperature exponent the DS values are indeed consistent. For the MF fits $\overline{q}_\infty < \sqrt{\overline{q^2}_\infty} < \overline{q}^{\infty}_{\max}$, as it should be. It is notable that fitting with a wrong zero temperature exponent may produce entirely inconsistent results. For instance with $y = 0.2$ one finds $\overline{q}_\infty > 1 > \overline{q}^{\infty}_{\max}$.

3.3 Groundstate energy

The energy is a self-averaging quantity, in contrast to the observables of the previous subsection. Consequently, the $L = 12$ lattices may contribute sufficiently accurate results and we rely now on all four data sets. To check that the self-averaging property is indeed consistent with our data, we have performed a fit of the form $L^3\sigma^2(e) = a_1 + a_2 L^{-3}$, and the corresponding goodness of fit $Q = 0.85$ is entirely satisfactory.

Most interesting are the FSS fits for the groundstate energy. With $L = 12$ included, the MF fit with 1/Volume corrections, $e_L = a_1 + a_2 L^{-3}$ is ruled out ($Q = 10^{-5}$), whereas the DS fit $e_L = a_1 + a_2 L^{-y}$ with $y = 0.72$ is well consistent. The DS groundstate energy estimate

$$e = -1.8389 \pm 0.0040 \qquad (12a)$$

is considerably lower than results reported in the literature, $e = -1.75$ [30], $e = -1.76 \pm$

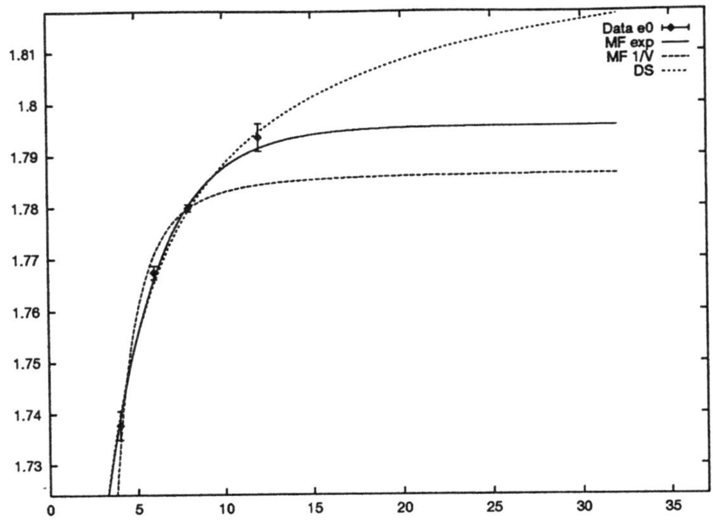

Figure 7: Fits for the groundstate energy (from up to down: DS, MF with exponentially small corrections, and MF with 1/Volume corrections).

0.02 [31] and $e = -1.7863 \pm 0.0028$ [27]. The latter discrepancy comes because our now very accurate data on lattices with $L \leq 8$ rule out the previously used fit with 1/Volume corrections. We like to remark that finding better (lower) groundstates energies for the $L = 12$ lattices (we noticed some problems with these estimates [28]), would only amplify the discrepancy with the MF fit, as the $L = 12$ groundstate energies would become even lower compared to the reliable other lattice sizes. This result is our strongest trend in favor of the DS and against the MF scenario. Unfortunately, there is still a catch to it. It may well be that the corrections to the uncritical MF behavior are exponentially small. Although FSS corrections for larger systems would then be greatly reduced, the disadvantage at the present level is that the appropriate $-e_L = -e + a_2 \exp(-a_3 L)$ fit has three free parameters. A consistent fit

$$-e = 1.7956 \pm 0.0042, \tag{12b}$$

$a_2 = -0.21 \pm 0.04$, $a_3 = 0.33 \pm 0.06$, and Q=0.19 is then still possible. All three fits are pictured in figure 7. The sharp turnover, necessary for the exponential fit looks rather unnatural. Again, accurate data on lattices up to $L = 16$ would unambiguously allow to differentiate this behavior from DS. But in contrast to results for the Parisi order parameters, it may be sufficient to simulate a fairly small number of lattices with $L \geq 12$.

4 Summary

The DS ansatz is so far consistent, and in particular also supported by results for the groundstate energy. Our investigation presents the first MC estimate of the zero temperature scaling exponent, which then enters self–consistently into a number of other estimates. Obviously, our lattices are too small to allow seminal results. In particular, the MF picture is still a valid alternative. It is clear that, either by brute computer power or by algorithmic improvements, simulations on larger lattices will become feasible. It seems, we are approaching a numerical conclusion about the correct groundstate picture of the 3*d* EAI model.

Acknowledgements: This work is supported, in part, by the Department of Energy, contracts DE-FG05-87ER40319 and DE-FC05-85ER250000.

References

[1] M. Mézard, G. Parisi, and M.A. Virasoro, *Spin Glass Theory and Beyond* (World Scientific, 1987).

[2] K.H. Fischer and J.A. Hertz, *Spin Glasses* (Cambridge University Press, 1991).

[3] D.L. Stein (editor), *Spin Glasses and Biology*, Directions in Condensed Matter Physics - Vol. 6, (World Scientific, 1992).

[4] R.N. Bhatt and A.P. Young, Phys. Rev. Lett. **54**, 924 (1985)

[5] A.T. Ogielski and I. Morgenstern, Phys. Rev. Lett. **54**, 928 (1985).

[6] A.T. Ogielski, Phys. Rev. B **32**, 7384 (1985).

[7] R.N. Bhatt and A.P. Young, Phys. Rev. B **37**, 5606 (1988).

[8] E. Marinari, G. Parisi and F. Ritort, preprint, cond-mat/9310041 (1993).

[9] W.L. McMillan, J. Phys. C **17**, 3179 (1984).

[10] A.J. Bray and M.A. Moore, J. Phys. C **18**, L699 (1985).

[11] D.S. Fisher and D.A. Huse, Phys. Rev. Lett. **56**, 1601 (1986).

[12] A.J. Bray and M.A. Moore, in *Heidelberg Colloquium on Glassy Dynamics*, edited by J.L. van Hemmen and I. Morgenstern, Lecture Notes in Physics, Vol. 275 (Springer, New York, 1987).

[13] D.S. Fisher and D.A. Huse, Phys. Rev. B **38**, 386 (1988).

[14] A. Bovier and J. Fröhlich, J. Stat. Phys. **44**, 347 (1986).

[15] S. Caracciolo, G. Parisi, S. Patarnello and N. Sourlas, J. Phys. France **51**, 1877 (1990).

[16] D.A. Fisher and D.A. Huse, J. Physique I France **1**, (1991) 621.

[17] N. Kawashima and N. Ito, J. Phys. Soc. Japan **62**, 435 (1993).

[18] R.R.P. Singh and D.A. Huse, J. Appl. Phys. **69**, 5225 (1991).

[19] R.E. Hetzel, R.N. Bhatt and R.R.P. Singh, Europhys. Lett. **22**, 383 (1993).

[20] B. Berg and T. Neuhaus, Phys. Lett. B **267**, 249 (1991).

[21] B. Berg and T. Neuhaus, Phys. Rev. Lett. **68**, 9 (1992).

[22] B. Berg and T. Celik, Phys. Rev. Lett. **69**, 2292 (1992); Int. J. Mod. Phys. C **3**, 1251 (1992).

[23] B. Berg, U. Hansmann, and T. Celik, preprint, submitted to Phys. Rev. B.

[24] G.M. Torrie and J.P. Valleau, J. Comp. Phys. **23**, 187 (1977).

[25] E. Marinari and G. Parisi, Europhys. Lett. **19**, 451 (1992).

[26] W. Kerler and P. Rehberg, preprint, cond-mat/9402049 (1994).

[27] B. Berg, T. Celik and U. Hansmann, Europhys. Lett. **22**, 63 (1993).

[28] B. Berg and C. Vohwinkel, in preparation.

[29] W.H. Press, B.P. Flannery, S.A. Teukolsky, and W.T. Vetterling, *Numerical Recipes* (Cambridge University Press, 1988).

[30] S. Kirkpatrick, Phys. Rev. B **16**, 4630 (1977).

[31] I. Morgenstern and K. Binder, Z. Phys. B **39**, 227 (1980).

Recent Developments in Monte-Carlo Simulations of First-Order Phase Transitions

W. Janke

Institut für Physik, Johannes Gutenberg-Universität, D-55099 Mainz, Germany

Abstract. In the past few years considerable progress has been made in Monte Carlo simulations of first-order phase transitions and in the analysis of the resulting finite-size data. In this paper special emphasis will be placed on multicanonical simulations using multigrid update techniques, on numerical estimates of interface tensions, and on accurate methods for determining the transition point and latent heat.

1 Introduction

First-order phase transitions play an important role in many fields of physics [1–3]. Well-known examples are field-driven transitions in magnets, crystal melting, the nematic-isotropic transition in liquid crystals or, at much higher energy scales, the deconfining transition in hot quark-gluon matter and the various transitions in the evolution of the early universe [3].

An important property of first-order phase transitions is phase coexistence. For field-driven transitions as, e.g., in the Ising model at low temperatures or the ϕ^4 theory discussed below, this is reflected in the canonical ensemble by a highly double-peaked probability distribution $P_{\rm can}(m)$ of the magnetization m. To sample in Monte Carlo (MC) simulations the two peaks with the correct relative weight the system has to pass many times through mixed phase configurations. For finite periodic systems of size L^d, such configurations contain (at least) two interfaces which contribute an additional term $2\sigma L^{d-1}$ to the free energy, where σ is the interface tension. Compared to the peak maxima they are hence suppressed by an additional Boltzmann factor $\propto \exp\{-2\sigma L^{d-1}\}$ which implies an exponential divergence of the autocorrelation time with system size, $\tau \propto \exp\{2\sigma L^{d-1}\}$. This property is sometimes called supercritical slowing down. The same arguments apply to temperature-driven first-order phase transitions as, e.g. in Potts models, where m has to be replaced by the energy e and $P_{\rm can}(m)$ by $P_{\rm can}(e)$.

Much effort has been spent in the past few years to develop efficient methods for numerical studies of this important class of phase transitions. Both, improved update schemes for data generation and refined techniques for data analysis have been studied. To overcome the slowing down problem, a so-called multicanonical method has been proposed which is basically a reweighting approach and can, in principle, be combined with any legitimate update algorithm. To date most applications employed the local heat-bath or Metropolis algorithms. These studies showed that (at least for all practical purposes) the exponential slowing down is indeed reduced to a much weaker power-law divergence with increasing system size. Since the remaining slowing down problem is, however, still severe, several studies have tried to combine the multicanonical approach with other update algorithms which are known to be much more efficient in the case of continuous phase transitions. In Sec. 2, after a brief review of the multicanonical method,

Springer Proceedings in Physics, Vol. 78

Computer Simulation Studies in Condensed-Matter Physics VII
Editors: D.P. Landau · K.K. Mon · H.-B. Schüttler

special emphasis will be laid on a recently proposed combination with multigrid update algorithms which was shown to give a further real-time improvement of about one order of magnitude for a two-dimensional ϕ^4 lattice model.

The uniform accuracy of the probability distribution in multicanonical simulations led to reinvestigations of a long-known histogram technique to determine the interface tension between the coexisting phases at the transition point. The 2D q-state Potts model, the 2D and 3D Ising model and the 2D ϕ^4 lattice model as well as the (3+1)D SU(3) lattice gauge theory have been studied so far. In Sec. 3.1 a summary is given of numerical results for the interface tension of 2D Potts models which can be compared with a recently derived analytical formula. This formula relies on an exact expression for the correlation length in the disordered phase at the transition point. Some direct numerical tests of this expression are presented in Sec. 3.2.

Parallel to the algorithmic developments many exact results concerning the finite-size scaling behavior at (strong) first-order phase transitions could be proven. This formulation suggested refined methods to estimate the transition point and latent heat from finite-size data, which are discussed in Sec. 3.3. The important point is that these estimates are exponentially close to the infinite-volume limit, i.e., they do not show the typical power-law corrections $\propto 1/L^d$ of the traditional observables and are hence of improved accuracy.

Finally, Sec. 4 contains a few concluding remarks.

2 Improved Generation of Monte Carlo Data

2.1 Multicanonical reweighting

The slowing down problem at first-order phase transitions is directly related to the double-peak shape of the canonical probability distribution $P_{\rm can}(m)$ or $P_{\rm can}(e)$. In multicanonical simulations [4, 5] of field-driven transitions this problem is avoided by simulating an auxiliary distribution $P_{\rm muca}(m) = P_{\rm can}(m) \times \exp(-f(m))$, where the reweighting factor $\exp(-f(m)) \equiv w(m)^{-1}$ is adjusted iteratively until $P_{\rm muca}(m) = const.$ between the two peaks [6–8]. This gives the mixed phase configurations the same statistical weight as the pure phases. Precisely the same idea applies to temperature-driven transitions with m replaced by e. Canonical expectation values $\langle \mathcal{O} \rangle_{\rm can}$ of any observable $\mathcal{O}$ can be recovered by the reweighting formula $\langle \mathcal{O} \rangle_{\rm can} = \langle w\mathcal{O} \rangle_{\rm muca} / \langle w \rangle_{\rm muca}$, where $\langle \dots \rangle_{\rm muca}$ denote expectation values with respect to the multicanonical distribution $P_{\rm muca}$.

Using local algorithms to update the multicanonical distribution it was demonstrated for various models [9–16] that the exponential slowing down is indeed reduced to a power-law behavior $\tau \propto V^\alpha = L^{d\alpha}$ with $\alpha \approx 1 - 1.5$, as one would expect from a simple random walk argument. While this is clearly an important step forward, the remaining slowing down problem is still severe. In fact, it is even worse than in simulations of critical phenomena where standard local MC algorithms yield $\tau \propto L^z$ with $z \approx 2$ [17, 18]. Here, however, several improved (mostly non-local) update algorithms (overrelaxation, cluster, multigrid, ...) are available which greatly reduce or even completely eliminate the slowing down problem [19–21].

In order to further improve the performance of Monte Carlo simulations of first-order phase transitions it is therefore quite natural to study combinations of multicanonical reweighting with these improved update algorithms. Overrelaxation techniques have been used in the context of SU(3) lattice gauge theory [16], but due to the complexity of this system no systematic investigation of the performance has been reported. Cluster updates are difficult to implement in a straightforward way since, due to the reweighting factor

$f(m)$ or $f(e)$, the multicanonical Hamiltonian is implicitly non-local. The more involved solution [11] proposed for Potts models yields a reduced exponent α and should thus be favorable for large systems. For moderate lattice sizes, however, it is again not clear how much is gained in real computer time. Multigrid update techniques finally can be quite efficient for problems with continuous fields as will be discussed in the next subsections.

2.2 Multigrid Monte Carlo Update Schemes

The general strategy of multigrid MC update schemes [22–24] is to perform collective moves on successively longer length scales in a systematic order as extensively discussed in the context of partial differential equations [25]. Both the type of collective moves and the sequence of length scales are parameters of multigrid schemes which can be defined in two equivalent ways: in a *unigrid* formulation where the update scheme is described in terms of the field variables on the original (fine) lattice, or in a *recursive multigrid* formulation where successively coarser lattices and associated Hamiltonians are introduced [19]. While the unigrid formulation is very transparent and easy to program, the more involved recursive multigrid formulation is numerically much more efficient.

In a unigrid formulation the collective update proposals are usually taken as square excitations, that is all fields in successively larger blocks of size $1, 2^d, 4^d, \ldots, V = 2^{nd}$ are moved by the same amount ϕ_b. Another possibility are excitations of the form of pyramids where ϕ_b is the amplitude of the central displacement. We then accept or reject the proposed collective move in accordance with the usual Metropolis formula. In a recursive multigrid formulation the block amplitudes ϕ_b are identified with field variables $\phi_i^{(k)}$ on coarse lattices $\Xi^{(k)}$ (with 2^{kd} sites), and the shape of the excitations is controlled by an operator $\mathcal{P}$ which interpolates the coarse grid fields back to the next finer grid $\Xi^{(k+1)}$. The square excitations correspond to a piecewise constant interpolation ($\mathcal{P}(\phi_1^{(k)}, \phi_2^{(k)}, \ldots) = (\phi_1^{(k)}, \phi_1^{(k)}, \phi_2^{(k)}, \phi_2^{(k)}, \ldots)$), and the pyramids correspond to a piecewise linear interpolation. To update the fields $\phi_i^{(k)}$ one defines a coarse grid Hamiltonian recursively as $H^{(k)}(\phi_i^{(k)}) = H^{(k+1)}(\phi_i^{(k+1)} + \mathcal{P}\phi_i^{(k)})$, $H^{(n)} = H$, and initializes the $\phi_i^{(k)}$ to zero. An update sweep over the coarse grid $\Xi^{(k)}$ then corresponds to updating the amplitudes ϕ_b of blocks of size $2^{(n-k)d}$ in the unigrid formulation. The multigrid formulation goes actually one step further and also defines the sequence of coarse grids or length scales in a recursive way. More precisely the update on level k thus consist of a) m_1 presweeps using any valid update algorithm with Hamiltonian $H^{(k)}$, b) calculating the Hamiltonian for the next coarser grid and initializing the field variables $\phi_i^{(k-1)}$ to zero. One then c) updates the field variables $\phi_i^{(k-1)}$ by applying the multigrid update γ_{k-1} times, d) interpolates the variables of level $k-1$ back to the finer grid, and e) performs another m_2 postsweeps on level k. On the coarsest grid $\Xi^{(0)}$ only steps a) and e) can be performed and the recursion stops. At step c) the sequence of coarse grids is defined by the cycle parameters γ_k. For $\gamma_k \equiv \gamma = 1$ (V-cycle) every grid is given the same weight, while for $\gamma_k \equiv \gamma = 2$ (W-cycle) the coarser grids are updated more frequently. The cycle names derive from their graphical representations which very much resemble the letters V and W; see Fig. 1.

For the Gaussian model it can be proven [19] that critical slowing down is completely eliminated a) for piecewise linear interpolation (or higher) and any cycle or b) for piecewise constant interpolation and the W-cycle (or higher). For a V-cycle with piecewise constant interpolation this has not yet been proven and, in fact, numerical evidence indicates that in this case $z = 1$ [26, 27].

An important feature of the recursive multigrid formulation is the fact that the computational labor W per cycle is proportional to the work W_0 per sweep of the original

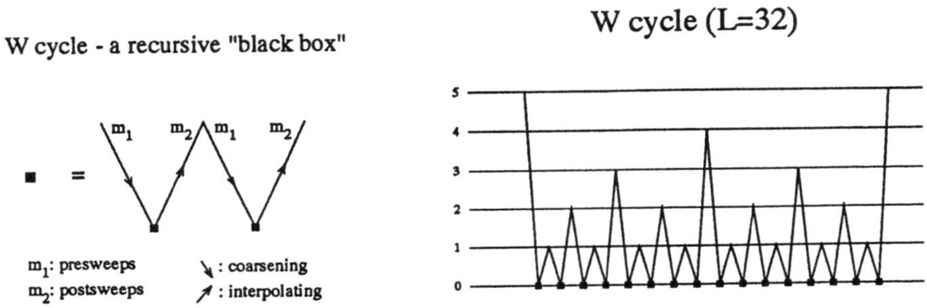

Figure 1: The recursive definition of the W-cycle and the expanded W-cycle for $L = 32$.

lattice, $W \approx W_0(1 - \gamma/2^d)^{-1}$ ($\gamma < 2^d$), independent of the lattice size. In the unigrid formulation, on the other hand, the estimates are $W \approx W_0 \log_2 L$ for the V-cycle and $W \approx W_0 L^{\log_2 \gamma}$ for $\gamma \geq 2$, i.e, $W \approx W_0 L$ for a W-cycle.

2.3 Multicanonical Multigrid Monte Carlo Method

When adapting multigrid schemes to multicanonical simulations one only has to make sure that also the reweighting factor $f(m)$ or $f(e)$ is treated properly [27–29]. Let us first consider the case of a field-driven transition in the unigrid viewpoint. Using piecewise constant interpolation, a proposed move ϕ_b for a block of size 2^{kd} would change the magnetization by $\Delta m = 2^{kd}\phi_b/V$. The decision of acceptance is thus simply to be based on the value of $\Delta E_{\text{muca}} = \Delta E_{\text{can}} + f(m + 2^{kd}\phi_b V) - f(m)$. In the recursive multigrid implementation the multicanonical modification is very similar since the coarse grid magnetization can be computed in the standard way. For a temperature-driven transition Δm would have to be replaced by $\Delta E_{\text{can}}/V$, which is also straightforward to implement in both the unigrid and multigrid formulation. For a discussion of more general situations, see Ref.[29].

2.4 Application to the 2D ϕ^4 Model

As an application of the multicanonical multigrid algorithm we considered in Ref. [29] the scalar ϕ^4 lattice field theory defined by the partition function

$$Z = \prod_{i=1}^{V} \left[\int_{-\infty}^{\infty} d\phi_i \right] \exp\left(-\sum_{i=1}^{V} \left(\frac{1}{2}(\vec{\nabla}\phi_i)^2 - \frac{\mu^2}{2}\phi_i^2 + g\phi_i^4 \right) \right), \tag{1}$$

with $g = 0.25$ and $\mu^2 > \mu_c^2(g) > 0$. Here $\mu_c^2(g)$ is a line of critical points separating the disordered ($\mu^2 < \mu_c^2(g)$) and ordered ($\mu^2 > \mu_c^2(g)$) phase. For $d \geq 2$ and $\mu^2 > \mu_c^2(g)$, the reflection symmetry is spontaneously broken in the infinite volume limit $L \to \infty$, and the average magnetization $\langle m \rangle = \langle (1/V) \sum_{i=1}^{V} \phi_i \rangle$ acquires a nonvanishing expectation value. Consequently, if a term $h \sum_i \phi_i$ is added to the energy, the system exhibits first-order phase transitions driven by the field h.

We have studied the field-driven first-order phase transition between the two ordered phases in $d = 2$ dimensions at $g = 0.25$ for $\mu^2 = 1.30, 1.35$, and 1.40. These points are sufficiently far away from the critical point at $\mu_c^2 = 1.265(5)$ [30] to display the typical behavior already on quite small lattices. A sensitive measure of the strength of the transition is the order-order interface tension σ_{oo} between the + and − phase, which turns out [29] to be $\sigma_{oo} = 0.03443(47), 0.09785(60)$, and $0.16577(73)$ [31].

To evaluate the performance of the multicanonical multigrid algorithm, we recorded the time series for several observables $\mathcal{O}$ and studied their autocorrelation functions $A(j) = \rho(j)/\rho(0)$, where $\rho(j) = \langle \mathcal{O}_i \mathcal{O}_{i+j} \rangle - \langle \mathcal{O}_i \rangle^2$. The exponential autocorrelation time τ^{exp} follows from its asymptotic decay for large j, $A(j) \propto \exp(-j/\tau^{\text{exp}})$, and the integrated autocorrelation time τ^{int} is obtained in the large k limit of $\tau(k) = 1/2 + \sum_{j=1}^{k} A(j)$. Here we shall concentrate only on the magnetization m which reflects most directly the (pseudo) dynamics of the transitions between the two ordered phases. In general we are interested in the variance ϵ^2 of the (weakly biased) estimator $\hat{\mathcal{O}} = \sum_{i=1}^{N_m} w(m_i)\mathcal{O}_i / \sum_{i=1}^{N_m} w(m_i) \equiv \overline{w_i \mathcal{O}_i}/\overline{w_i}$ for a canonical expectation value $\langle \mathcal{O} \rangle_{\text{can}}$. To facilitate a direct comparison with canonical simulations, we *define* for multicanonical simulations an effective autocorrelation time τ^{eff} by the standard error formula for N_m correlated measurements,

$$\epsilon^2 = \sigma_{\text{can}}^2 2\tau^{\text{eff}}/N_m, \tag{2}$$

where $\sigma_{\text{can}}^2 = \langle \mathcal{O}_i^2 \rangle_{\text{can}} - \langle \mathcal{O}_i \rangle_{\text{can}}^2$ is the variance of single measurements in the canonical distribution. The squared error ϵ^2 can be estimated either directly by jack-knife blocking procedures [32], or by applying standard error propagation to $\hat{\mathcal{O}} = \overline{w_i \mathcal{O}_i}/\overline{w_i}$, which involves the (multicanonical) variances and covariances of $w_i \mathcal{O}_i$ and w_i, and the three associated integrated autocorrelation times $\tau^{\text{int}}_{\mathcal{O};\mathcal{O}} \equiv \tau^{\text{int}}_{\mathcal{O}}$, $\tau^{\text{int}}_{w\mathcal{O};w\mathcal{O}} \equiv \tau^{\text{int}}_{w\mathcal{O}}$, and $\tau^{\text{int}}_{w\mathcal{O};\mathcal{O}} = \tau^{\text{int}}_{\mathcal{O};w\mathcal{O}}$ [28, 29].

Some results for $\mu^2 = 1.30$ from runs with at least 20 000 τ^{int}_{wm} Metropolis sweeps or W-cycles (using piecewise constant interpolation without postsweeps) are collected in Table 1. We see that τ^{exp}_m and τ^{int}_m agree well with each other, indicating that the corresponding autocorrelation function $A(j)$ can be approximated by a single exponential. While the τ^{exp}_{wm} agree with τ^{exp}_m within error bars, the integrated autocorrelation times τ^{int}_{wm} are significantly smaller. This implies that for this observable $A(j)$ is composed of many different modes. We further observe that the difference between τ^{int}_{wm} and τ^{eff}_m can be quite appreciable, reflecting the varying probability distribution shapes with increasing L. For comparison with previous work we have also included the diffusion times τ^{flip}_m, which are defined as one quarter of the time it takes to travel from one (canonical) peak to the other and back.

The log-log plot in Fig. 2 shows the behavior of τ^{eff}_m as a function of the lattice size L. Least-squares fits of τ^{eff}_m to a power law, $\tau^{\text{eff}}_m \propto V^{\alpha}$, yield for both update algorithms exponents of $\alpha \approx 1.2, 1.4$, and 1.5 for $\mu^2 = 1.30, 1.35$ and 1.40, i.e., the multigrid update does not improve the asymptotic behavior. The effective autocorrelation times of the

Table 1: Autocorrelation times in units of sweeps resp. cycles for the Metropolis (M) and multigrid W-cycle (W) update in multicanonical simulations of the model (1) in $d = 2$ with $g = 0.25$ and $\mu^2 = 1.30$.

	$L=8$		$L=16$		$L=32$		$L=64$
	M	W	M	W	M	W	W
τ^{exp}_m	212(12)	11.30(32)	668(23)	37.2(2.0)	3120(200)	148(11)	746(62)
τ^{int}_m	204.4(4.0)	10.88(12)	690(11)	34.69(76)	2984(63)	150.0(4.0)	758(37)
τ^{exp}_{wm}	209(12)	11.34(33)	655(31)	36.9(2.0)	2880(190)	146(13)	600(120)
τ^{int}_{wm}	171.1(3.4)	9.82(11)	509.8(8.9)	27.58(59)	1840(40)	96.6(2.4)	374(23)
τ^{eff}_m	322.7(6.1)	18.51(20)	1258(21)	67.4(1.3)	6050(120)	321.9(7.6)	1724(86)
τ^{flip}_m	463.5(6.4)	30.82(25)	1759(24)	91.7(1.3)	7780(140)	428.2(8.9)	1922(85)

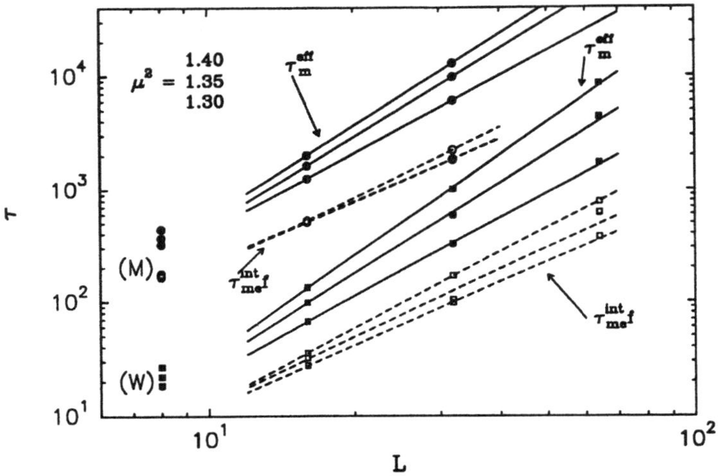

Figure 2: Effective autocorrelation times τ_m^{eff} for the model (1) in $d = 2$ with $g = 0.25$ as a function of lattice size L.

W-cycle, however, are about 20 times smaller than those of the Metropolis algorithm. If we finally take into account that a W-cycle requires more elementary operations than a Metropolis sweep, we obtain with our implementations on a CRAY Y-MP a *real time* improvement factor of about 10.

3 Refined Data Analysis

3.1 Interface Tension

A quantity of central importance for the kinetics of first-order phase transitions is the interface tension σ between the coexisting phases [1, 2]. As discussed earlier, on finite periodic lattices of size L^d, this is reflected by a double-peak structure of the probability distribution for the energy or magnetization, with the minimum between the two peaks dominated by mixed phase configurations with two interfaces contributing an excess free energy of $2\sigma L^{d-1}$. This suggests [43] to extract the interface tension from the infinite volume limit of

$$2\sigma^{(L)} = \frac{1}{L^{d-1}} \ln(P_{\text{max}}^{(L)} / P_{\text{min}}^{(L)}). \tag{3}$$

For accurate results the system has to travel many times between the two peaks, which is a serious problem in canonical simulations of strong first-order transitions (large σ). The multicanonical sampling, on the other hand, is just designed for this purpose since it gives the same relative errors for $P_{\text{max}}^{(L)}$ and $P_{\text{min}}^{(L)}$ and thus optimizes the error on $\sigma^{(L)}$.

Using this so-called histogram method interface tensions have been estimated at the temperature driven first-order phase transitions of 2D q-state Potts models [9–12, 14, 51] and $N_t = 2$ SU(3) lattice gauge theory [10, 16], and at the field-driven transitions of the 2D and 3D Ising [15] and the 2D ϕ^4 model [29]. For the 2D Ising model good agreement with the exact Onsager formula was obtained. For the 2D q-state Potts model with

Table 2: Comparison of analytical and numerical results for the order-disorder interface tension $2\sigma_{od}$ in 2D q-state Potts models.

q	ξ_d	$2\sigma_{od}$	$2\sigma_{od}$ (MC)	
7	48.095907	0.020792	0.0241(10)	Janke *et al.* [10]
			0.02348(38)	Rummukainen [11]
			0.0228(24)	Grossmann and Gupta [12]
8	23.878204	0.041879	0.045	Janke [51]
10	10.559519	0.094701	0.09781(75)	Berg and Neuhaus [9]
			0.10	Janke [51]
			0.0950(5)	Billoire *et al.* [14]
15	4.180954	0.239179	0.263(9)	Gupta [42]
20	2.695502	0.370988	0.3714(13)	Billoire *et al.* [14]

Hamiltonian [34]

$$H = -J \sum_{\langle ij \rangle} \delta_{s_i s_j}; \quad s_i \in \{1, \ldots, q\}, \tag{4}$$

some numerical results are compiled in Table 2 and compared with an analytical formula [35] (derived *after* the first numerical results were already published), which follows by combining (a) the assumption of complete wetting [36, 37] with (b) duality arguments [38] and (c) an exact result for the correlation length $\xi_d(\beta_t)$ in the disordered phase at the transition point $\beta_t = \ln(1 + \sqrt{q})$ [39] (see also [40, 41]),

$$2\sigma_{od} = \sigma_{oo} = 1/\xi_d = \frac{1}{4} \sum_{n=0}^{\infty} \ln \left[\frac{1 + w_n}{1 - w_n} \right], \tag{5}$$

where $w_n = \left[\sqrt{2} \cosh\left((n + \frac{1}{2})\pi^2/2v\right)\right]^{-1}$ with $v = \ln\left(\frac{1}{2}\left[\sqrt{\sqrt{q}+2} + \sqrt{\sqrt{q}-2}\right]\right)$. More precisely, it was shown [36] that $2\sigma_{od} \le \sigma_{oo}$ for *all* $q \ge 5$. The opposite inequality could only be derived for $q > q_0$ (with $4 < q_0 < \infty$ being a sufficiently large constant), but by basic thermodynamic arguments it is commonly believed that it actually also holds for all $q \ge 5$.

Overall the numerical and analytical values are in good agreement, but noteworthy is the systematic trend of the numerical data to overestimate the analytical values, which are actually exact *upper* bounds.

3.2 Correlation Length

To test the formula (5) for the correlation length $\xi_d(\beta_t)$ directly, we performed further Monte Carlo simulations [44] of the Potts model (4) and measured the $k_y = 0$ projection g of the correlation function

$$G(i,j) = \langle \delta_{s_i s_j} - 1/q \rangle, \tag{6}$$

at β_t in the disordered phase. Here we simply choose the lattice sizes large enough in order to suppress transitions to the ordered phase. Estimates of autocorrelation times showed [44] that in this situation the optimal update procedure consists of many single-cluster [45] update steps combined with one multiple-cluster [46] update for efficient measurements using the improved estimator

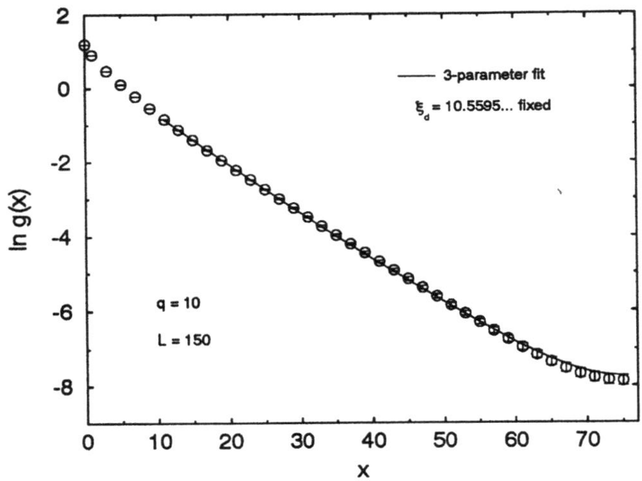

Figure 3: Semi-log plot of the correlation function $g(x)$ vs distance x for $q = 10$. The solid line is a three-parameter fit with ξ_d held fixed at its theoretical value.

$$G(i,j) = \frac{q-1}{q}\langle\Theta(i,j)\rangle, \tag{7}$$

where $\Theta(i,j)=1$, if i and j belong to the same cluster, and $\Theta=0$ otherwise.

The numerical data for $q = 10$ at β_t are shown in Fig. 3. From the curvature of $g(x)$ for small distances it is clear that fits with the simplest Ansatz for periodic boundary conditions, $g(x) = a\cosh((x - L/2)/\xi_d)$, will only work for very large x. We therefore tried to use the more general Ansatz $g(x)=a\cosh((x - L/2)/\xi_d) + b\cosh(c(x - L/2)\xi_d)$, with four parameters a, b, c, and ξ_d. The solid line in Fig. 3 shows a fit where ξ_d is held fixed at its theoretical value and only the three remaining parameters are optimized. The good quality of the fit over a wide x range shows that our data are compatible with the theoretical expectations. To get a more quantitative measure we next performed four-parameter fits with ξ_d as a *free* parameter. Fits over the same x range yield an estimate of $\xi_d(\beta_t) = 9.2(8)$, which is about 15% smaller than the exact value but within the statistical errors still consistent. By restricting the fit interval to larger x values, we observed a tendency to higher values but with the drawback of increased error bars. Since our data for $q = 15$ and 20 exhibit the same qualitative behavior, we believe that higher excitations cannot be neglected at the distances studied so far ($x_{\max} = L/2 \approx 7\xi_d$). To cope with this problem we are currently performing further simulations on $2L \times L$ lattices, which should allow to study the correlations over even larger distances.

3.3 Accurate Determination of the Transition Point and Latent Heat

The traditional way to locate first-order transition points is based on the scaling behavior of the specific-heat maxima $C_{\max}(V)$ and Binder-parameter minima $B_{\min}(V)$, where $C(V,\beta) = \beta^2 V(\langle e^2\rangle - \langle e\rangle^2)$ and $B(V,\beta) = 1 - \langle e^4\rangle/3\langle e^2\rangle^2$. In the idealized infinite volume limit, these quantities develop singularities at the transition point β_t. In a finite volume $V = L^d$ the singularities are smoothed out and, depending on its strength, the transition

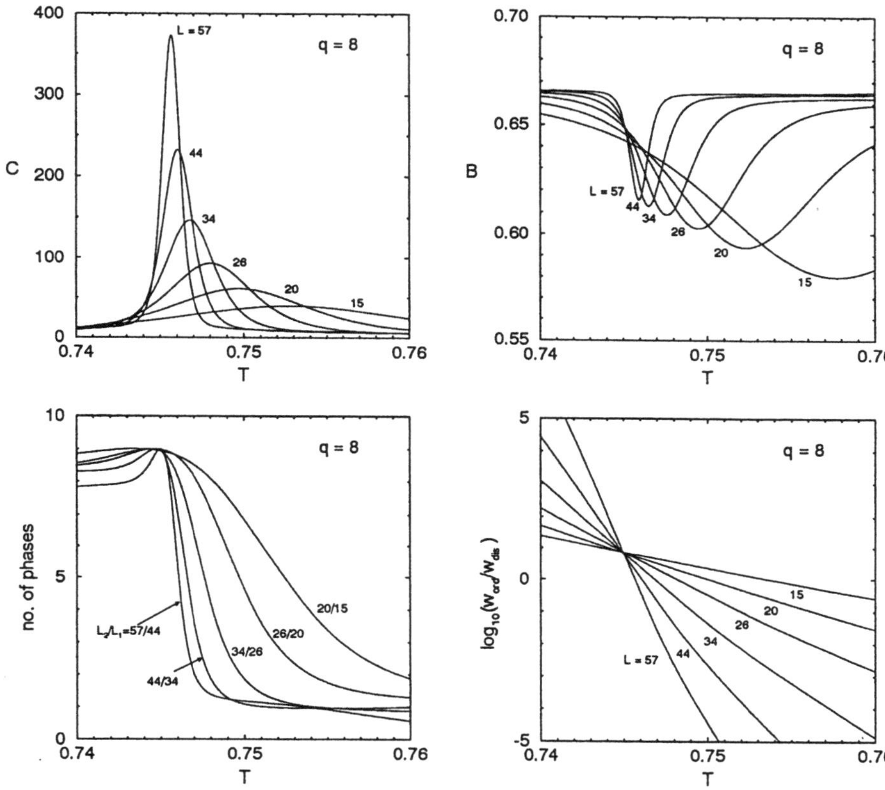

Figure 4: Finite-size scaling of the traditional observables, specific heat C and Binder parameter B, compared with the observables defined in eqs. (10), (12) for the two-dimensional 8-state Potts model.

is signalized by more or less pronounced finite peaks of the specific heat or dips of the Binder parameter near the infinite volume transition point. For an illustration see Fig. 4 where data for the two-dimensional 8-state Potts model are shown. If the volume is cubic or nearly cubic the location of the extrema is typically shifted by an amount $O(V^{-1})$ with respect to the actual infinite volume transition point and one may try to estimate $\beta_t \equiv 1/T_t$ from the finite volume results by extrapolations in $1/V$ [47].

In addition to random statistical errors the data are in general also systematically shifted by exponential corrections which are difficult to take into account theoretically. This makes the extrapolation of finite volume data not always reliable. It is therefore desirable to find definitions of a finite volume transition point which involve *no* power-law corrections at all.

The starting point for such definitions is the observation that on lattices with periodic boundary conditions the partition function of a model describing the coexistence of one disordered and q ordered phases can be written for large enough q as [48, 49]

$$Z_{\text{per}}(V,\beta) = \left(\sum_{m=0}^{q} e^{-\beta f_m(\beta) V}\right)\left(1 + \mathcal{O}\left(V e^{-L/L_0}\right)\right). \qquad (8)$$

Here $L_0 < \infty$ is a constant, L is the linear length of the lattice, and $f_m(\beta)$ is the metastable free energy density of the phase m. It can be defined in such a way that it is equal to the idealized infinite volume free energy density $f(\beta)$ if m is stable and strictly larger than $f(\beta)$ if m is unstable [48, 49]. This implies (see also Refs.[50, 51]) that in the infinite volume limit the parameter

$$N(V,\beta) = Z_{\text{per}}(V,\beta)e^{\beta f(\beta)V} \tag{9}$$

is equal to the number of stable phases at the inverse temperature β, i.e., $N(\beta) = \lim_{V\longrightarrow\infty} N(V,\beta) = q$ in the ordered phase, $N(\beta) = q+1$ at the transition point β_t, and $N(\beta) = 1$ in the disordered phase. A natural definition of a finite volume transition point $\beta_t(V)$ is thus the point where $N(V,\beta)$ is maximal. Due to the bound (8) (and similar bounds for derivatives [48, 49]) this definition leads to only *exponentially* small shifts of $\beta_t(V)$ with respect to the infinite volume transition point β_t.

The free energy $f(\beta)$ in (9) is only defined in the thermodynamic limit and hence not accessible to numerical simulations. It is therefore necessary to eliminate this term by, e.g., forming a suitable ratio. Instead of (9) one may look for the maximum of the *number-of-phases* parameter

$$N(V_1,V_2,\beta) = \left[\frac{Z_{\text{per}}(V_1,\beta)^\alpha}{Z_{\text{per}}(V_2,\beta)}\right]^{\frac{1}{\alpha-1}}, \tag{10}$$

where $\alpha = V_2/V_1$ [50, 51]. By inserting (8) it is straightforward to verify that with increasing temperature $N(V_1,V_2,\beta)$ smoothly interpolates between the values q, $q+1$, and 1. Actual simulation results for $\alpha = V_2/V_1 \approx 1.6$ are shown in Fig. 4. The locations of the maxima define the desired, only exponentially shifted finite-volume transition points $\beta_t(V_1,V_2)$ which will be denoted $\beta_{V/V}$. By differentiating $\ln N$ with respect to β one readily sees that determining $\beta_{V/V}$ amounts to solving $\alpha E(V_1,\beta_{V/V}) = E(V_2,\beta_{V/V})$ or $e(V_1,\beta_{V/V}) = e(V_2,\beta_{V/V})$, i.e., to locating the crossing point of the internal energies per site, $e \equiv E/V$, of the two lattices of different size.

The numerical determination of $\beta_{V/V}$ requires simulations on two different lattices. In Ref.[50] we have therefore proposed another definition of a finite-volume transition point which requires data from one lattice only. Its definition is based on the fact that at the infinite-volume transition point all free energies $f_m(\beta)$ are equal, so that eq. (8) implies

$$w_o \equiv \sum_{m=1}^{q} e^{-\beta_t f_m(\beta_t)V} = qe^{-\beta_t f_d(\beta_t)V} \equiv qw_d, \tag{11}$$

apart from exponentially small corrections. Here the free energy density of the "zeroth", disordered phase is denoted by f_d, and $w_{o,d}$ are the associated statistical weights of the coexisting phases. A natural definition of a finite-volume transition point β_W is thus the point where the ratio of the total weight of the q ordered phases to the weight of the disordered phase approaches q. More precisely we introduce the *ratio-of-weights* parameter

$$R(V,\beta) \equiv W_o/W_d \equiv \sum_{E<E_0} P_{V,\beta}(E)/\sum_{E\geq E_0} P_{V,\beta}(E), \tag{12}$$

where $P_{V,\beta}(E)$ are the (double-peaked) energy histograms, and determine β_W by solving

$$R(V,\beta_W) = q. \tag{13}$$

The parameter E_0 in (12) is defined by reweighting [52] the energy distribution to the temperature where the two peaks of $P_{V,\beta}(E)$ have equal height and then taking E_0 as

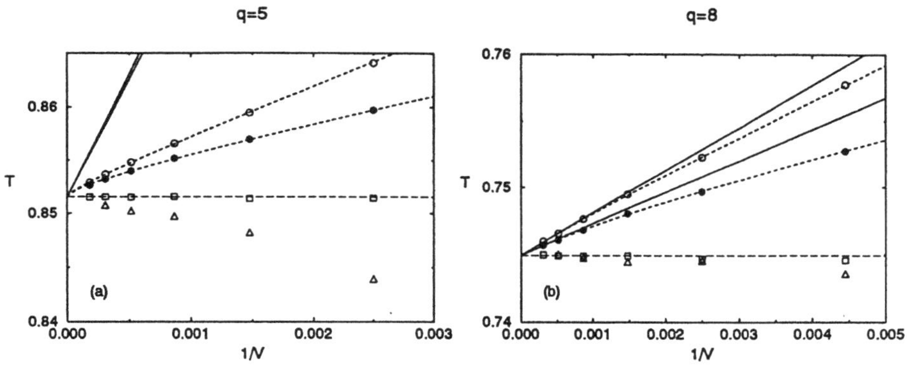

Figure 5: Scaling behavior of the finite-volume transition points defined by the Binder-parameter minimum (○), the specific-heat maximum (•), the maximum of the number-of-phases parameter (△), and the ratio-of-weight parameter (□). The solid straight lines are the exactly known $1/V$ corrections corresponding to ○ and •, and the dashed, almost interpolating curves show exponential fits (including the $1/V$ corrections) to these data. Note that the $(1/V)^2$ corrections are almost invisible on this scale and in any case point in the "wrong" upward direction. The long dashed horizontal lines indicate the exact infinite-volume transition points.

the energy at the minimum between the two peaks. Other definitions of E_0 would be reasonable as well, as for example the internal energy at the temperature where the specific heat is maximal. Since it is expected that the relative height of the minimum between the two peaks decreases like $\exp(-2\sigma L^{d-1})$ as $L \to \infty$, all these definitions do in fact only differ by exponentially small corrections and it is a matter of practical convenience to choose E_0. In Fig. 4 the logarithm of the ratio-of-weights parameter $R(V,\beta)$ is plotted as a function of temperature.

In (13) we have assumed that the number of ordered phases, q, is known by general arguments. If this is not the case, one may use the crossing points $\beta_{W/W}$ satisfying $R\left(V_1, \beta_{W/W}\right) = R\left(V_2, \beta_{W/W}\right)$ as estimates for β_0. The value of R at the crossing point then gives the ratio of the number of coexisting ordered and disordered phases. This, however, requires again the simulation on two lattices of different size.

The convergence properties of the four finite-volume transition points are compared in Fig. 5 for a very weak ($q = 5$) and a rather strong ($q = 8$) first-order transition. Notice the surprisingly fast convergence of $\beta_W(V)$ even for $q = 5$.

The ratio-of-weights method leads naturally to a finite-volume definition of the latent heat [51] which also should have only exponentially small corrections with respect to the infinite volume limit. Since

$$\ln(w_o/w_d) = -\beta V(f_o - f_d), \tag{14}$$

the slopes of $R(V,\beta)$ in Fig. 4 at the crossing point may be used to define

$$\Delta e(V) = e_d(V) - e_o(V) = \frac{d}{d\beta}\ln(W_o/W_d)/V = -\frac{1}{T^2}\frac{d}{dT}\ln(W_o/W_d)/V. \tag{15}$$

The resulting estimates $\Delta e(V)$ are plotted in Fig. 6 and compared with the traditional

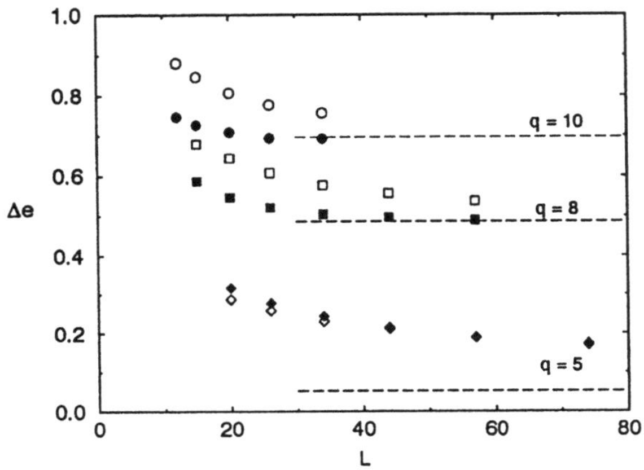

Figure 6: The finite-volume latent heat Δe of the 2D q-state Potts model vs linear lattice size L. The open symbols show the traditional estimates from the peak locations of $P_{V,\beta}(E)$, and the filled symbols follow from the slopes of the ratio-of-weight parameter. The dashed horizontal lines show the exactly known infinite-volume limits [54].

definition based on the peak locations of $P_{V,\beta}(E)$ [53]. For strong first-order transitions ($q = 8$ and 10) the asymptotic limit is indeed reached much faster with the new definition. For a very weak transition ($q = 5$), on the other hand, both methods yield comparable estimates which are still far away from the limiting value.

4 Concluding Remarks

The material reviewed in these lecture notes is, admittedly, a somewhat biased selection of recent developments in this field. Different improved Monte Carlo update schemes are discussed, e.g., in Refs.[55, 56]. Also, refined data analyses have been developed along many different lines. For a recent review focusing on the analysis of probability distributions and various cumulants, e.g., see Ref.[57].

The combination of multicanonical reweighting with multigrid update schemes leads to a significant reduction of autocorrelation times by a roughly constant factor. It still remains a challenge to develop Monte Carlo algorithms which exhibit no slowing down at a first-order phase transition.

Acknowledgments

I would like to thank Bernd Berg, Christian Borgs, Stefan Kappler, Mohammad Katoot and Tilman Sauer for fruitful collaborations and the DFG for a Heisenberg fellowship.

References

[1] J.D. Gunton, M.S. Miguel, and P.S. Sahni, in *Phase Transitions and Critical Phenomena*, Vol. 8, eds. C. Domb and J.L. Lebowitz (Academic Press, New York, 1983).

[2] K. Binder, Rep. Prog. Phys. **50** (1987) 783.

[3] For recent reviews, see, e.g., *Dynamics of First Order Phase Transitions*, eds. H.J. Herrmann, W. Janke, and F. Karsch (World Scientific, Singapore, 1992).

[4] B.A. Berg and T. Neuhaus, Phys. Lett. **B267** (1991) 249.

[5] For earlier related ideas, see G.M. Torrie and J.P. Valleau, Chem. Phys. Lett. **28** (1974) 578; J. Comp. Phys. **23** (1977) 187.

[6] For a review, see B.A. Berg, in Ref.[3], p.311; and in *Physics Computing '92*, eds. R.A. deGroot and J. Nadrchal (World Scientific, Singapore, 1993), p.23.

[7] W. Janke in Ref.[3], p.365; and in *Physics Computing '92*, eds. R.A. deGroot and J. Nadrchal (World Scientific, Singapore, 1993), p.351.

[8] A.D. Kennedy, in *Lattice '92*, Nucl. Phys. (Proc. Suppl.) **30** (1993) 96.

[9] B.A. Berg and T. Neuhaus, Phys. Rev. Lett. **68** (1992) 9.

[10] W. Janke, B.A. Berg, and M. Katoot, Nucl. Phys. **B382** (1992) 649.

[11] K. Rummukainen, Nucl. Phys. **B390** (1993) 621.

[12] B. Grossmann and S. Gupta, Phys. Lett. **B319** (1993) 215.

[13] A. Billoire, T. Neuhaus, and B.A. Berg, Nucl. Phys. **B396** (1993) 779.

[14] A. Billoire, T. Neuhaus, and B.A. Berg, Nucl. Phys. **B413** (1994) 795.

[15] B.A. Berg, U. Hansmann, and T. Neuhaus, Phys. Rev. **B47** (1993) 497; Z. Phys. **B90** (1993) 229.

[16] B. Grossmann and M.L. Laursen, in Ref.[3], p.375; and Nucl. Phys. **B408** (1993) 637; B. Grossmann, M.L. Laursen, T. Trappenberg, and U.-J. Wiese, Phys. Lett. **B293** (1992) 175.

[17] K. Binder, in *Monte Carlo Methods in Statistical Physics*, ed. K. Binder (Springer, New York, 1979), p.1.

[18] See also the articles in *Finite-Size Scaling and Numerical Simulations of Statistical Systems*, ed. V. Privman (World Scientific, Singapore, 1990).

[19] A.D. Sokal, *Monte Carlo Methods in Statistical Mechanics: Foundations and New Algorithms*, Cours de Troisième Cycle de la Physique en Suisse Romande, Lausanne, 1989; and in *Quantum Fields on the Computer*, ed. M. Creutz (World Scientific, Singapore, 1992), p.211.

[20] C.F. Baillie, Int. J. Mod. Phys. **C1** (1990) 91.

[21] R.H. Swendsen, J.-S. Wang, and A.M. Ferrenberg, in *The Monte Carlo Method in Condensed Matter Physics*, ed. K. Binder (Springer, Berlin, 1992), p.75.

[22] J. Goodman and A.D. Sokal, Phys. Rev. Lett. **56** (1986) 1015; Phys. Rev. **D40** (1989) 2035.

[23] G. Mack, in *Nonperturbative quantum field theory*, Cargèse 1987, ed. G.'t Hooft et al. (Plenum, New York, 1988); G. Mack and S. Meyer, Nucl. Phys. **B** (Proc. Suppl.) **17** (1990) 293; M. Hasenbusch, S. Meyer, and G. Mack, Nucl. Phys. **B** (Proc. Suppl.) **20** (1991) 110.

[24] D. Kandel, E. Domany, D. Ron, A. Brandt, and E. Loh, Jr., Phys. Rev. Lett. **60** (1988) 1591; D. Kandel, E. Domany, and A. Brandt, Phys. Rev. **B40** (1989) 330.

[25] W. Hackbusch, *Multi-Grid Methods and Applications* (Springer, Berlin, 1985); S.F. McCormick (ed.), *Multigrid Methods. Theory, Applications, and Supercomputing* (Dekker, New York, 1988).

[26] W. Janke and T. Sauer, *Chem. Phys. Lett.* **201** (1993) 499.

[27] W. Janke and T. Sauer, in Proceedings of the Fourth International Conference *Path Integrals from meV to MeV: Tutzing '92*, eds. H. Grabert, A. Inomata, L.S. Schulman, and U. Weiss, (World Scientific, Singapore, 1993), p.17.

[28] W. Janke and T. Sauer, FU Berlin preprint FUB-HEP 9/93, to appear in Phys. Rev. **E49** (March 1994).

[29] W. Janke and T. Sauer, FU Berlin preprint FUB-HEP 15/93, submitted to J. Stat. Phys.

[30] R. Toral and A. Chakrabarti, Phys. Rev. **B42** (1990) 2445. See also A. Milchev, D.W. Heermann, and K. Binder, J. Stat. Phys. **44** (1986) 749.

[31] These values are comparable with the more familiar order-disorder interface tensions in 2D q-state Potts models with $q = 9, 14$, and 19; cp. Table 2.

[32] R.G. Miller, *Biometrika* **61** (1974) 1; B. Efron, *The Jackknife, the Bootstrap and other Resampling Plans* (SIAM, Philadelphia, PA, 1982).

[33] K. Binder, Phys. Rev. **A25** (1982) 1699; Z. Phys. **B43** (1981) 119.

[34] For a general review of the Potts model, see e.g. F.Y. Wu, Rev. Mod. Phys. **54** (1982) 235; *ibid.* **55** (1983) 315 (*Erratum*).

[35] C. Borgs and W. Janke, J. Phys. I (France) **2** (1992) 2011.

[36] J. de Coninck, A. Messager, S. Miracle-Solé, and J. Ruiz, J. Stat. Phys. **52** (1988) 45; R. Schonmann, J. Stat. Phys. **52** (1988) 61.

[37] A. Messager, S. Miracle-Solé, J. Ruiz, and S. Shlosman, Commun. Math. Phys. **140** (1991) 275.

[38] L. Laanait, Phys. Lett. **A124** (1987) 480.

[39] E. Buffenoir and S. Wallon, J. Phys. **A26** (1993) 3045.

[40] A. Klümper, A. Schadschneider, and J. Zittartz, Z. Phys. **B76** (1989) 247.

[41] A. Klümper, Int. J. Mod. Phys. **B4** (1990) 871.

[42] S. Gupta, preprints HLRZ 22/93, 65/93, Jülich (1993).

[43] K. Binder, Phys. Rev. **A25** (1982) 1699; Z. Phys. **B43** (1981) 119.

[44] W. Janke and S. Kappler, Mainz preprint (1993), to appear in *Lattice '93*, Nucl. Phys. (Proc. Suppl.) **B** (1994); and preprint in preparation.

[45] U. Wolff, Phys. Rev. Lett. **62** (1989) 361.

[46] R.H. Swendsen and J.S. Wang, Phys. Rev. Lett. **58** (1987) 86.

[47] In general, the exponent may depend on the model and the type of boundary conditions, see M. E. Fisher and A.N. Berker, Phys. Rev. **B26** (1982) 2507; V. Privman and M.E. Fisher, J. Stat. Phys. **33** (1983) 385; K. Binder and D.P. Landau, Phys. Rev. **B30** (1984) 1477; M.S.S. Challa, D.P. Landau, and K. Binder, Phys. Rev. **B34** (1986) 1841; P. Peczak and D.P. Landau, Phys. Rev. **B39** (1989) 11932; V. Privman and J. Rudnik, J. Stat. Phys. **60** (1990) 551. For a review, see V. Privman in Ref.[18]. A special case with $\alpha = 2$, e.g., happens if the number of stable phases on both sides of the transition are equal, see Refs. [48, 49].

[48] C. Borgs and R. Kotecky, J. Stat. Phys. **61** (1990) 79; Phys. Rev. Lett. **68** (1992) 1734.

[49] C. Borgs, R. Kotecky, and S. Miracle-Solé, J. Stat. Phys. **62** (1991) 529.

[50] C. Borgs and W. Janke, Phys. Rev. Lett. **68** (1992) 1738.

[51] W. Janke, Phys. Rev. **B47** (1993) 14757.

[52] A.M. Ferrenberg and R.H. Swendsen, Phys. Rev. Lett. **61** (1988) 2635; *ibid.* **63** (1989) 1658 (*Erratum*).

[53] J. Lee and J.M. Kosterlitz, Phys. Rev. Lett. **65** (1990) 137; Phys. Rev. **B43** (1991) 3265.

[54] R.J. Baxter, J. Phys. **C6** (1973) L445.

[55] A. Hüller, Z. Phys. **B88** (1992) 79; R.W. Gerling and A. Hüller, Z. Phys. **B90** (1993) 207.

[56] W. Kerler and A. Weber, Phys. Rev. **B47** (1993) 11563.

[57] K. Binder, K. Vollmayr, H.-P. Deutsch, J.D. Reger, M. Scheucher, and D.P. Landau, in Ref.[3], p.253.

Statistical Dependence and Related Topics

M. Kikuchi[1], *N. Ito*[2], *and Y. Okabe*[2]

[1]Department of Physics, Osaka University, Toyonaka 560, Japan
[2]Department of Applied Physics, The University of Tokyo, Bunkyo-ku, Tokyo 113, Japan
[3]Department of Physics, Tokyo Metropolitan University, Hachioji, Tokyo 192-03, Japan

Abstract On the basis of the dynamical interpretation of Monte Carlo simulations, we discuss the relation of the equilibrium relaxation time, the susceptibility and the statistical error. We introduce a new quantity called *the statistical dependence time* τ_{dep}, which gives the reduction factor for the statistical degree of freedom due to the dynamical correlations between the data. A new method is proposed for calculating equilibrium relaxation time using τ_{dep}, the method which does not require knowledge of any time-displaced correlation function. We apply this method to the critical dynamics of Ising models in two and three dimensions, and estimate the dynamical critical exponent z precisely. Systematic errors in response functions due to short simulations are also discussed from the viewpoint of the statistical dependence.

1 Introduction

Although Monte Carlo methods are stochastic methods, sequences obtained by simulations are often regarded as realistic dynamical sequences of a model in concern; And several dynamical properties, both in equilibrium and in non-equilibrium, are studied using Monte Carlo simulations. Dynamical correlations in equilibrium are characterized by the equilibrium relaxation time τ. Any correlations, both spatial and temporal, in a physical quantity can be investigated using the appropriate correlation functions. In fact, the relaxation time τ is usually estimated from the asymptotic behavior of the corresponding time-displaced correlation function. On the other hand, existence of the characteristic time scale τ means that the measurements made in a simulation are not statistically independent of each other. The statistical dependence here is thus a different aspect of the dynamical correlation. According to above considerations, we expect that alternative approaches for studying dynamics are possible, which may be called *methods of statistical dependence analysis.*

Recently, we have proposed a new method for calculating τ, based on the statistical dependence analysis.[1] This method no longer requires calculation of time-displaced correlation functions; Rather, it uses only

Springer Proceedings in Physics, Vol. 78
Computer Simulation Studies in Condensed-Matter Physics VII
Editors: D.P. Landau · K.K. Mon · H.-B. Schüttler

a ratio of equilibrium averages of the susceptibility and the statistical error of a thermodynamic quantity in concern. As a result, it enables us to make statistical analyses in an unambiguous manner; We can simply follow the standard analysing methods used for static quantities. It is in contrast with the conventional method, where unambiguous statistical analyses are not easy to be done, since the statistical errors in the time-displaced correlation function for different time are usually not independent of each other. In fact, if all the data in a time-displaced correlation function were simply regarded as independent ones, the resulting statistical error in τ would be systematically underestimated.

The aim of the present paper is to review recent progress in the statistical dependence method for studying dynamics. The topics we will deal with are the estimation of the dynamical critical exponents of the Ising models in two and three dimensions, and the reexamination of unbiased estimators for response functions. Before proceeding to these topics, we will describe the statistical dependence method in the next section.

2 The Statistical Dependence Time

Let us start from discussing the relation of the response function (*e.g.* magnetic susceptibility for magnetic models), the relaxation time, and the statistical error. Roughly and intuitively speaking, the square of the statistical error for a thermodynamic quantity is proportional to the corresponding susceptibility divided by τ. That relation was explored in their pioneering work by Müller-Krumbhaar and Binder[2] back in 1973, although in a different context. As will be shown later, however, the expression they obtained was valid only in the long-time limit. In this section, we will give an expression which is correct in any range of time. The new method for calculating τ will be proposed in the course of the discussion.

Suppose we make N independent simulations of a system under the same parameters (*e.g.* temperature), with measurements of a physical quantity Q made in constant time intervals; This interval is used as a unit of time in the following. Number of such measurements in each run is n. Under this situation, we can define two different kinds of (thermal) averages as follows:

1. The average over the measurements in a single run,

$$\langle Q\rangle_\alpha \equiv \frac{1}{n}\sum_{k=1}^{n} Q_\alpha(k), \tag{1}$$

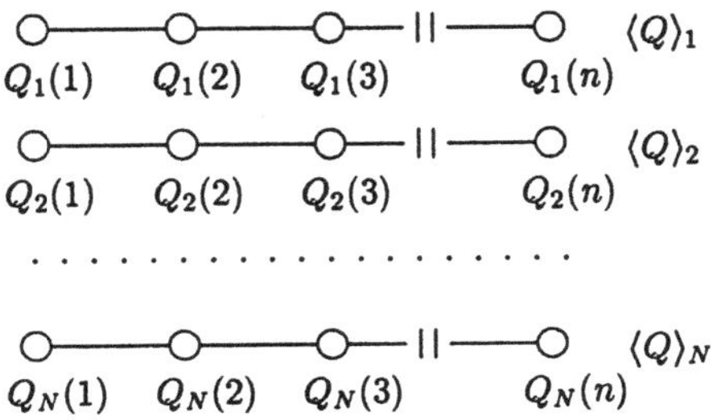

Figure 1: Schematic illustration of the simulation setup. Each simulation runs horizontally. The circles represent the measurements of a quantity Q; n measurements are made in each run. The average $\langle Q\rangle_\alpha$ is calculated from all the measurements in α -th run. The above procedure is repeated N times independently.

2. The average over the independent runs,

$$\overline{\langle Q\rangle} \equiv \frac{1}{N}\sum_{\alpha=1}^{N}\langle Q\rangle_\alpha, \tag{2}$$

where $Q_\alpha(k)$ denotes the value of the quantity Q at the k-th measurement in the α-th run, where $k = 1, 2, \cdots, n$ and $\alpha = 1, 2, \cdots, N$. The situation is schematically illustrated in Fig. 1.

It is well known in elementary statistics that if all the n data are statistically independent, the statistical degrees of freedom (DOF) is n, and the expectation value for the variance of $\langle Q\rangle$ (we omitted the suffix α here, since the expectation value does not depend on it) is

$$\mathrm{E}\left\{(\langle Q\rangle - \mu)^2\right\} = \frac{\sigma^2}{n}, \tag{3}$$

where $\mu \equiv \mathrm{E}\{Q\}$ and $\sigma^2 \equiv \mathrm{E}\left\{Q^2\right\} - \mu^2$ are the expectation values of Q and of the variance of Q, respectively. On the other hand, if the data are correlated to each other, as is usually the case in dynamical simulations, rather than they are independent, the DOF will be reduced due to the mutual correlations. Such reduction in the DOF can be discussed in terms of the time-displaced correlation function of Q. In fact, the expectation value for the variance of $\langle Q\rangle$ for the correlated data is given as

$$\mathrm{E}\left\{(\langle Q\rangle - \mu)^2\right\} = \frac{1}{n^2}\sum_{k,l=1}^{n}\left(\mathrm{E}\{m(k)m(l)\} - \mu^2\right) = \frac{1}{n^2}\sum_{k,l=1}^{n}\mathrm{E}\{C(k,l)\}, \tag{4}$$

where $C(k,l)$ is the unnormalized time-displaced correlation function of the fluctuation in Q for "time" k and l. Let us assume that the relaxation of this correlation function is described by a single exponential function, as is often the case, with the associated relaxation time τ; That is, we assume that the correlation function is expressed as

$$\mathrm{E}\left\{C(k,l)\right\} = \sigma^2 e^{-|k-l|/\tau}. \tag{5}$$

Then we get

$$\mathrm{E}\left\{(\langle Q\rangle - \mu)^2\right\} = \frac{\sigma^2}{n^2} \sum_{k,l=1}^{n} e^{-|k-l|/\tau}. \tag{6}$$

If we approximate the summation in this equation by an integration and take the integration range as $[0, \infty)$, we have

$$\mathrm{E}\left\{(\langle Q\rangle - \mu)^2\right\} = \frac{2\tau}{n}\sigma^2. \tag{7}$$

The order of the thermal average and the long-time limit can not be interchanged here, because $\lim_{n\to\infty}\langle Q\rangle = \mu$ by definition. Equation (7) is just the expression obtained by Müller-Krumbhaar and Binder. It should be noted that the limit $1/\tau \to 0$ is implicitly assumed in the integral approximation as well as the explicit long-time limit; Equation (7) is thus valid only in this limit.

Let us proceed further, without assuming the long-time limit. The summation in Eq. (6) can be calculated explicitly, and gives

$$\sum_{k,l=1}^{n} e^{-|k-l|/\tau} = n\left[\frac{1+\Lambda}{1-\Lambda} - \frac{2\Lambda(1-\Lambda^n)}{n(1-\Lambda)^2}\right], \tag{8}$$

where $\Lambda \equiv \exp(-1/\tau)$. Then Eq. (6) can be expressed in a similar form as Eq. (3) and (7) as follows:

$$\mathrm{E}\left\{(\langle Q\rangle - \mu)^2\right\} = \frac{2\tau_{\mathrm{dep}}}{n}\sigma^2, \tag{9}$$

with

$$\tau_{\mathrm{dep}} = \frac{1}{2}\left[\frac{1+\Lambda}{1-\Lambda} - \frac{2\Lambda(1-\Lambda^n)}{n(1-\Lambda)^2}\right]. \tag{10}$$

We introduced a quantity τ_{dep} having the dimension of time; We call it *the statistical dependence time.* Comparing Eq. (9) with Eq. (3), we see that the factor $n/2\tau_{\mathrm{dep}}$ plays a role of the DOF for correlated data; The DOF is thus reduced by a factor $1/2\tau_{\mathrm{dep}}$ due to the dynamical correlations. In other words, only $n/2\tau_{\mathrm{dep}}$ among n measurements have statistical significance, and $2\tau_{\mathrm{dep}}$ is interpreted as the mean interval of successive statistically significant measurements. It should be noted that this reduction factor $1/2\tau_{\mathrm{dep}}$ depends not only on τ but also on n, while in the conventional argument based on Eq. (7) the reduction factor is just $1/2\tau$ irrespective of n.

The statistical dependence time τ_{dep} can be estimated by simulations using Eq. (9) as follows: The unbiased estimator $(\delta Q)^2$ for $\mathrm{E}\left\{(\langle Q\rangle - \mu)^2\right\}$ is calculated from N independent runs as

$$(\delta Q)^2 = \frac{N}{N-1}\left(\overline{\langle Q\rangle^2} - \overline{\langle Q\rangle}^2\right). \tag{11}$$

The quantity $|\delta Q|/\sqrt{N}$ is commonly used as the estimator for the statistical error in Q. The variance σ^2, on the other hand, is calculated as the fluctuation χ_Q, which is physically proportional to the response function associated to Q:

$$\chi_Q = \overline{\langle Q^2\rangle} - \overline{\langle Q\rangle}^2, \tag{12}$$

Combining these equations, we get our final expression for the estimator of τ_{dep}:

$$\tau_{\text{dep}} = \frac{n(\delta Q)^2}{2\chi_Q} = \frac{nN}{2(N-1)}\frac{\overline{\langle Q\rangle^2} - \overline{\langle Q\rangle}^2}{\overline{\langle Q^2\rangle} - \overline{\langle Q\rangle}^2}. \tag{13}$$

Therefore, what is proportional to the square of the statistical error is not τ but τ_{dep}. Once τ_{dep} is thus estimated, we can then get τ by solving Eq. (10).

The above discussions can be generalized to cases of multi-exponential relaxation,

$$\mathrm{E}\left\{C(k,l)\right\} = \sum_I a_I e^{-|k-l|/\tau_I}, \tag{14}$$

where τ_I is the relaxation time of the I-th mode, and a_I the corresponding amplitude. Then the statistical dependence time becomes

$$\tau_{\text{dep}} = \sum_I \frac{a_I}{2}\left[\frac{1+\Lambda_I}{1-\Lambda_I} - \frac{2\Lambda_I(1-\Lambda_I^n)}{n(1-\Lambda_I)^2}\right], \tag{15}$$

where $\lambda_I = \exp(-1/\tau_I)$. For practical use, however, it is convenient to follow the same procedure as in the case of single-exponential relaxation; That is, we calculate the *effective* relaxation time τ_{eff} by solving Eq. (10) instead of Eq. (15). The effective relaxation time τ_{eff} thus estimated can, in fact, be used as a substitute for the so-called integrated relaxation time τ_{int}, whenever the integrated relaxation time has a definite meaning. Thus it obeys the same scaling law as τ_{int} at criticality. More detailed discussions are given in Ref. [1].

3 Dynamical Critical Exponent of 3D Ising Model

In this section, we present the results of applying the method described in the previous section to the critical dynamics of the three-dimensional Ising model. The Hamiltonian is given as follows:

$$H = -J \sum_{\langle i,j \rangle} \sigma_i \sigma_j, \quad (\sigma = \pm 1) \tag{16}$$

where the interactions act on the nearest-neighbor spin pairs. At the critical point, the relaxation time obeys the finite-size scaling law with the dynamical critical exponent z:

$$\tau \sim L^z, \tag{17}$$

where L is the linear dimension of the system. Therefore, by calculating τ for several system sizes at the critical point, we can estimate z.

We performed Monte Carlo simulations for the Ising model on the simple-cubic lattices of the size $L \times L \times (L+1)$, with the skewed boundary conditions imposed in the L–directions, and the periodic boundary condition in the $(L+1)$–direction. We employed the Metropolis transition probability. The Simulations were fully vectorized by means of the two-interpenetrating-sublattice updating procedure. A multispin coding technique were also used, which assign a spin to a single bit. The Simulations were made at the critical point $J/T_c = 0.221654$, with 64 independent runs made simultaneously by means of the recycling algorithm.[3] We took 19 different system sizes, among which the smallest one was $L = 3$, and the largest one was $L = 91$.

We measured the magnetization and its square by the simulations, and calculated the associated τ_{eff} from τ_{dep} by solving Eq. (10). Length of the simulations were adjusted to be about $10\tau_{\text{dep}}$, with the same length discarded for equilibration. The number of the independent runs N was taken from 23×64 at smallest to 1400×64 at largest, depending on the system size. These procedures were repeated several times for estimating the statistical errors in τ_{eff}. Figure 2 shows the log-log plot of τ_{eff} against the *average* linear dimension $L_{\text{av}} \equiv \left[L^2(L+1)\right]^{1/3}$. As can be seen in the figure, the data fit very well to a straight line. But we found by detailed statistical analysis that τ_{eff} for $L = 3$ deviated significantly from the straight line. This deviation can be seen clearly in Fig. 3 where we show z calculated from eight successive data; That is, we took a set of eight successive data points in Fig. 2, and fitted them to a straight line; The values of z estimated that way for all the possible sets are shown in the figure against the mid value L_{mid} of L_{av} for each set. The error bars indicate 1σ of the fit. We see in the figure that only z estimated from the set including $L = 3$ gives an extraordinarily large value, while all the others fall in the range 2.030 ± 0.005. We confirmed also from the χ^2 value

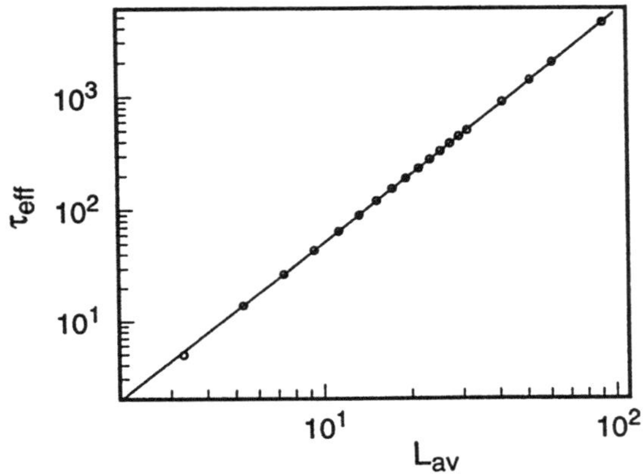

Figure 2: Log-log plot of τ_{eff} against $L_{\text{av}} \equiv [L^2(L+1)]^{1/3}$ for 3D Ising model. The errors are much smaller than the symbols. The straight line represents $z = 2.030$.

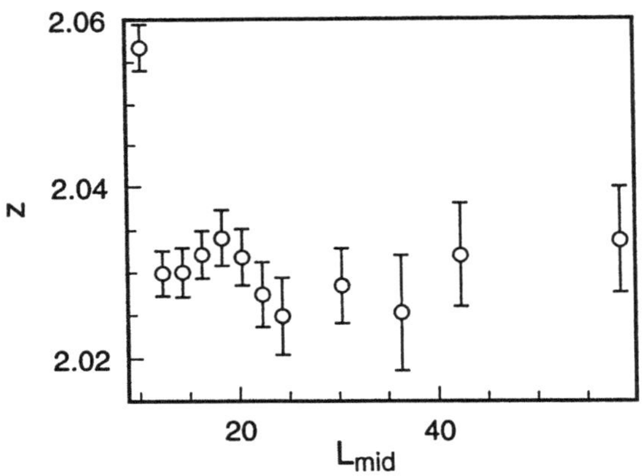

Figure 3: The value of z calculated from the data sets, each of which consists of eight successive sizes. The horizontal axis shows the mid value of L_{av} for each set.

of the fit that only τ_{eff} for $L = 3$ does not fit to the straight line. Based on these statistical observations, we can safely exclude $L = 3$ from further analyses. The least-square fit of the rest 18 data gave $z = 2.030 \pm 0.004$; Here, we took 2σ of the fit as the error instead of 1σ, just for the sake of safety. The corresponding χ^2/DOF of the fit was 1.05; Thus the fitting of the data to the straight line is extremely good. Besides the statistical error, there may be the error due to uncertainty in T_c;[4] Taking that error into account, our final estimate is $z = 2.03 \pm 0.01$, which is still quite accurate one.

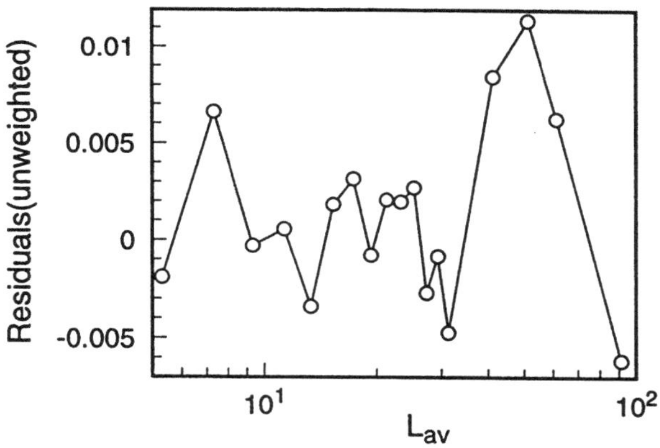

Figure 4: The residuals of the fit for 3D Ising model against the system size. The values shown are not weighted by the statistical errors.

Since the dynamics of three-dimensional models are equivalent to the statics of some four-dimensional models, which will be highly anisotropic in the fourth dimension, it is sometimes expected that the dynamical scaling law for three-dimensional models has a logarithmic correction. We, however, have not found any indication of the logarithmic correction to the dynamical finite-size scaling law, Eq. (17) within the present accuracy; Indeed, no particular tendency can be seen in the residuals of the fit, which we plot in Fig. 4 against the system size.

4 Dynamical Critical Exponent of 2D Ising Model

Dynamical critical exponent of two-dimensional Ising Model was also calculated in the same line as in the previous section. We made simulations of square-lattice Ising model of the size $L(L+1)$ with the skewed boundary condition in L–direction and the periodic boundary condition in another direction. Again we calculated the magnetization and its square by the simulations at the critical point $J/T_c = 0.4406868$, and estimated τ_{eff}. The same method of simulations as in the previous section was used. The smallest system size was $L = 11$, and the largest one was $L = 201$. Again the length of the simulations were adjusted to be about $10\tau_{\text{dep}}$, with the same length discarded for equilibration. The number of the independent runs N was taken from 64 at smallest to 4000×64 at largest, depending on the system size.

Figure 5 shows the log-log plot of τ_{eff} against $L_{\text{av}} \equiv [L(L+1)]^{1/2}$. Again, the data seem to fit very well to a straight line. We follow sim-

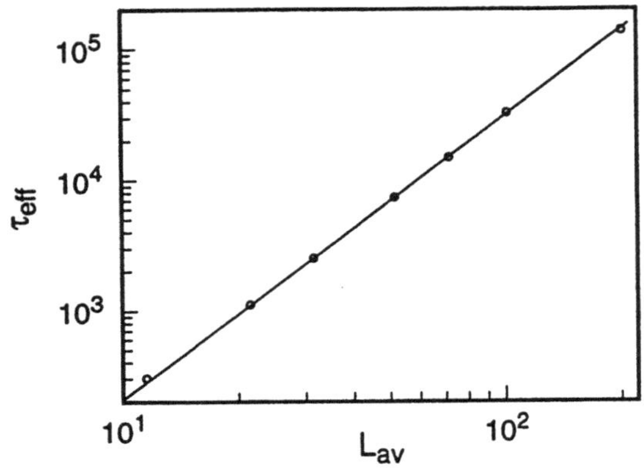

Figure 5: Log-log plot of τ_{eff} against $L_{\text{av}} \equiv [L(L+1)]^{1/2}$ for 2D Ising model. The errors are much smaller than the symbols. The straight line represents $z = 2.173$.

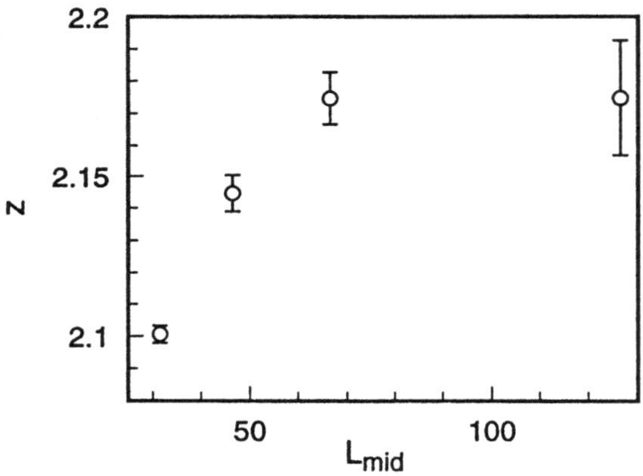

Figure 6: The value of z calculated from the data sets, each consisting of four successive sizes against the mid value of L_{av} for each set.

ilar statistical analyses made in the previous section. Figure 6 shows z calculated from the data sets of four successive sizes (not eight, this time) against L_{mid}. Unlike the three-dimensional case, the value of z approaches slowly to its asymptotic value. Even so, we can regard that the five data from $L = 31$ to $L = 201$ are in the asymptotic region as is seen in Fig. 6. We confirmed that also from χ^2/DOF of the fit. By the least-square fit of these five data to the finite-size scaling form Eq. 17, we estimated $z = 2.173 \pm 0.016$, again with the error of 2σ. The value of χ^2/DOF of the fit is 0.86; Thus the data are fitted very well to a straight line.

5 On unbiased estimators of response functions

In this section, we reexamine the arguments given by Ferrenberg, Landau, and Binder[7] on systematic errors in response function measurements, from the viewpoint of the statistical dependence time. Unlike the previous sections, we consider the case where only a single run of a simulation is made. It is well known that the unbiased estimator for the variance, that is, the second-order cumulant of Q from n *independent* measurements is given as

$$\chi_Q = \frac{n}{n-1}\left(\langle Q\rangle^2 - \langle Q\rangle^2\right). \tag{18}$$

The factor $n/(n-1)$ appears here for correction of the bias on the estimate due to the finite number of the measurements. This correction factor has already been taken into account in Eq. (11) in Sec. 1. From the statistical-physics points of view, the variance χ_Q is interpreted as the response function associated to Q. Therefore, the estimators for response functions calculated from the fluctuations in general should also be corrected by the above factor; Otherwise, the measured response functions will be systematically underestimated. In actual simulations, however, the measurements are not independent of each other, as has been discussed so far. Therefore, we can not simply take $n/(n-1)$ as the correction factor; The correction factor should be larger than it, due to the dynamical correlations.

The above systematic errors due to finite-length simulations have been pointed out in ref. [7] They argued that the response function estimated from n measurements $\chi_Q(n)$ and its true expectation value $\chi_Q(\infty)$ is related as

$$\chi_Q(n) = \chi_Q(\infty)\left(1 - \frac{2\tau}{n}\right). \tag{19}$$

Near the critical point, on the other hand, several response functions exhibit scaling behaviors. For example, the magnetic susceptibility χ_m at the critical point is scaled by the system size as

$$\chi_m(\infty) \sim L^{\gamma/\nu}, \tag{20}$$

where γ and ν are the critical exponents of the susceptibility and the correlation length, respectively. Combining these two scaling relations, Eq. (19) and Eq. (20), they proposed a brand new scaling relation, which may be called the finite-size finite-measurement scaling, for the susceptibility as

$$\chi_m(n) = aL^{\gamma/\nu}\left(1 - \frac{2\tau}{n}\right). \tag{21}$$

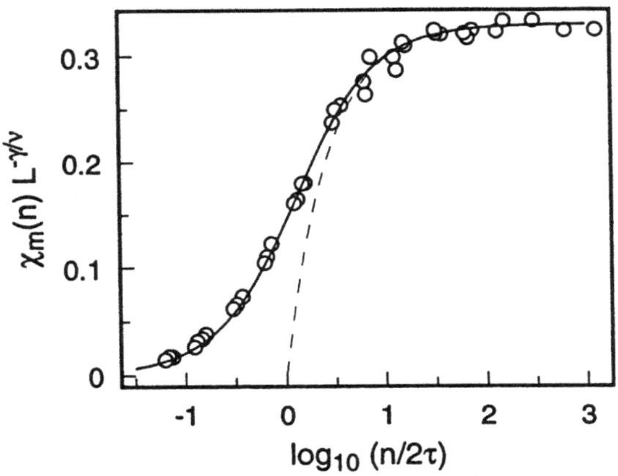

Figure 7: The finite-size finite-measurement scaling plot for the magnetic susceptibility of the 3D Ising model at the critical point. The symbols are taken from the simulations by Ferrenberg *et al.*;[7] The data for several sizes from $L = 16$ to 96 are plotted. The solid line represents the function $a(1 - 2\tau_{\mathrm{dep}}/n)$, while the dashed line represents the function $a(1 - 2\tau/n)$.

However, as we have discussed in Sec. 2, the statistical DOF for the correlated measurements are not $n/2\tau$ but $n/2\tau_{\mathrm{dep}}$ in general. Therefore, the scaling relation Eq. (21) should be corrected as follows:

$$\chi_m(n) = aL^{\gamma/\nu}\left(1 - \frac{2\tau_{\mathrm{dep}}}{n}\right). \tag{22}$$

Equation (22) is expected be valid in general, while Eq. (21) is valid only when $n/2\tau$ is sufficiently large.

In Fig. 7 we show the scaling plot of $\chi_m(n)$ for the three-dimensional Ising model at $J/T_c = 0.221654$; The symbols represent the data taken from the simulations in ref. [7]; The two scaling functions $1-2\tau/n$ and $1-2\tau_{\mathrm{dep}}/n$ are also shown by the dashed line and the solid line, respectively, with only the overall factor a being the fitting parameter. We see clearly that the scaling function $1 - 2\tau_{\mathrm{dep}}/n$ agrees very well to the simulation data. On the other hand, the scaling function $1-2\tau/n$ coincides with the data only for large $n/2\tau$, as is expected. It should be noted that since the effective DOF $n/2\tau_{\mathrm{dep}}$ depends not only on τ but also on n, so does the scaling function $1 - 2\tau_{\mathrm{dep}}/n$. We, however, found that as long as n is much larger than 10τ, the scaling function does not depends on n in the range of $n/2\tau$ shown in the figure.

6 Conclusions

So far we have discussed the recent progress in the statistical dependence method for studying dynamics. In the discussions in Sec. 2, we did not use any special property inherent in Monte Carlo simulations; Rather they are quite general, so that they can apply also to other sorts of dynamical simulations, such as the molecular dynamics method.

In Secs. 3 and 4, we have applied the statistical dependence time method for estimating the dynamical critical exponent z. Although the relaxations at the critical point are in general multi-exponential ones, we found that the present method worked very well. The dynamical critical exponents z for the Ising model both in two and three dimensions have been calculated by a lot of authors so far.[4] In case of three-dimensional Ising model, the estimates using the equilibrium correlation time spread within the range $1.99 \sim 2.17$. Among them, recent two estimates based on large-scale Monte Carlo simulations have given mutually incompatible values: 2.04 ± 0.03 by Wansleben and Landau[5] and 2.10 ± 0.02 by Heuer.[6] Our estimation agrees well with the former and disagrees with the latter within the range of error. Moreover, the statistical error in z in the present result is much smaller than that of the others. It should be noted that we did not use any special technique for estimating the statistical error in z; We just followed the standard methods used for estimating the errors in static quantities. On the other hand, the error estimation must be more or less complicated and thus necessarily ambiguous in the conventional methods. In case of the two-dimensional Ising model, the estimates of z so far are also scattering in a wide range. The statistical error quoted for the present result is one of the smallest errors among them.

In Sec. 4 we reexamined the argument on the systematic errors in response functions due to short simulations by Ferrenberg *et al.*. We presented the correct scaling function for the scaling relation they proposed. Consequently, we confirmed both the finite-size finite-measurement scaling they have proposed and our proposal that $n/2\tau_{\mathrm{dep}}$ gives the statistical DOF for correlated data, at the same time. It should be noted the following: In case of the magnetic susceptibility, or response functions associated to odd quantities of the order parameter in general, such systematic errors do not cast any serious problem, because one can calculate χ_m as just the square of m at $T \geq T_c$ (there can be no unique definition of χ_m at $T < T_c$ for finite systems). On the other hand, for the response functions of even quantities, such as the specific heat, one always has to be careful about that systematic errors; Even then, however, one can reduce the errors by making many independent runs and using Eq. (12).

Acknowledgements: Two of the authors (M. K. and N. I.) would like to thank Professor K. Binder and Professor D. Stauffer for their kind hospitality, fruitful discussions, and encouragement. Thanks are also due to Dr. H. Takano and Dr. T. Koma for valuable discussions. The work was partially supported by a Grant-in-Aid for Scientific Research on Priority Areas, "Computational Physics as a New Frontier in Condensed Matter Research", from the Ministry of Education, Science and Culture, Japan.

References

[1] M. Kikuchi and N. Ito: J. Phys. Soc. Jpn. **62** (1993) 3052.

[2] H. Müller-Krumbhaar and K. Binder: J. Stat. Phys. **8** (1973) 1.

[3] N. Ito, M. Kikuchi, and Y. Okabe: Int. J. Mod. Phys. **C4** (1993) 569.

[4] N. Ito: Physica **A192** (1993) 604 (also ref. [1]) and references therein for 3D
N. Ito: Physica **A196** (1993) 591 and references therein for 2D.

[5] S. Wansleben and D. P. Landau: Phys. Rev. **B43** (1991) 6006.

[6] H. O. Heuer: J. Phys. **A25** (1992) L567.

[7] A. M. Ferrenberg, D. P. Landau, and K. Binder: J. Stat. Phys. **63** (1991) 867.

Dynamical Scaling of the Island-Size Distribution and Percolation in Submonolayer Molecular Beam Epitaxy

*J.G. Amar, F. Family, and P.M. Lam**

Department of Physics, Emory University, Atlanta, GA 30322, USA
*Present address: Department of Physics, Southern University, Baton Rouge, LA 70813, USA

Abstract. The results of a detailed study of the evolution, scaling and percolation of islands in a model of submonolayer molecular-beam epitaxy appropriate for the case of dendritic island growth are presented. The scaling of the island density, monomer density, island-size distribution, structure factor, and pair-correlation function are studied as a function of the coverage θ and the ratio $R = D/F$ of the diffusion rate D to the deposition rate F. Our results span the full range of coverage starting from very low coverage all the way through the coalescence and percolation regimes. For small coverage the islands are fractal while at higher coverages they become compact, similar to what has been observed in Au/Ru(0001). For large R, four distinct scaling regimes are found: a low-coverage nucleation regime, an intermediate-coverage regime, an aggregation regime, and a coalescence and percolation regime. The scaling behavior in the first three regimes is compared with results obtained using a generalised rate equation approach. A novel "fractal" scaling form for the structure factor in the aggregation regime is derived. The dependence of the percolation threshold on R is also studied and found to show interesting non-monotonic behavior.

1. Introduction

Recent experimental work using scanning tunneling microscopy (STM) as well as diffraction techniques has enabled the detailed study of the submonolayer island morphology, density, and size distribution for a variety of systems ranging from homoepitaxial systems [1-4] to heteroepitaxial systems such as Pb/Cu(001) [5], and Au/Ru(0001) [6]. In addition, a variety of experiments have been performed (1,3-4) which attempt to measure the parameters of surface diffusion (activation energy for example) by measuring the scaling of the island density as function of temperature at low coverage. If one assumes that for fixed coverage θ, the island density $N(\theta)$ scales as $N = C_1(D/F)^{-\chi}$ where D is the diffusion rate in unit of hops per unit time and F is the deposition rate in units of monolayers deposited per unit time, then the exponent χ may be experimentally determined by studying the variation of N with F. Once χ is known the activation energy E_a for diffusion may be determined from the dependence of the island density as a function of temperature, while if the constant of proportionality C_1 is known the absolute diffusion rate may be de-

Springer Proceedings in Physics, Vol. 78
Computer Simulation Studies in Condensed-Matter Physics VII
Editors: D.P. Landau · K.K. Mon · H.-B. Schüttler

termined. Thus, the knowledge of the scaling behavior of the island density and distribution for simple models of island formation both in the very-early-time and the aggregation regime may be used to determine important physical quantities in epitaxial growth. In addition, the study of the island morphology and distribution may be useful in determining such quantities as the critical island size, the activation energy for edge diffusion, and a variety of bond strengths.

2. Simulations

In our model, adatoms (monomers) are deposited randomly on a square lattice with deposition rate F (in units of atoms per site per unit time) and then diffuse with diffusion rate D (nearest-neighbor hops per adatom per unit time). An adatom deposited on top of an existing island or monomer (second-layer growth) also diffuses with diffusion rate D. When an adatom encounters another adatom as its nearest neighbor, both atoms are 'frozen' (stop diffusing) and form a stable two-atom cluster or island. Similarly, when a diffusing adatom encounters an existing island, it is attached to the island cluster and becomes immobile. At each stage, either a deposition step (consisting of the deposition of one adatom at a randomly chosen site) or a diffusion step (consisting of the diffusion by one nearest-neighbor lattice-site of a randomly chosen walker) is carried out. In order to properly keep track of the competing rates of diffusion and deposition, the deposition step was carried out with probability $p_F = 1/(1 + N_1(D/F))$, where N_1 is the monomer density (per site). Similarly, the diffusion step was carried out with probability $p_D = 1 - p_F = N_1(D/F)/(1 + N_1(D/F))$. A list of all monomers or walkers was continually updated, and if a monomer encountered another particle (i.e. another monomer or part of a cluster) as its nearest-neighbor, then that monomer was added to the cluster and removed from the walker list.

We note that in our model there is no additional barrier or activation energy for jumping down a step (Schwoebel effect [18]). This is consistent with the observation for island growth of Au on Ru(0001) that Au atoms can easily diffuse over first layer island steps to condense at their edge in the bottom layer [6]. Thus, if a monomer deposited on an island encounters another adatom or island on the same adlayer it is frozen, while if it first diffuses over the edge of the island to the layer below it is added to the island. Thus for large D/F (which corresponds to the scaling regime) there is a negligible amount of second-layer growth (at least below percolation) since the second-layer atoms quickly diffuse to the edge of an island and down to the first layer where they join the cluster.

Our simulations typically involved system sizes ($L \times L$) ranging from $L =$ 400 to $L = 1024$, with periodic boundary conditions. (Smaller system sizes were also used to obtain snapshots of the island morphology.) Averages were taken over 50-200 runs and simulations were conducted for different values of the ratio $R = D/F$ ranging from 10 to 10^9. In our simulations, we kept track of the monomer density N_1, the total density N of first-layer islands, and the first-layer island-size distribution N_s, all as a function of the coverage θ from very low coverage ($\theta \simeq 10^{-5}$) to a coverage $\theta = 1$. In addition, we monitored the number of nucleation events in order to directly determine the rate of coalescence of islands as a function of coverage. We also calculated the normalized circularly-averaged first-layer density-density pair-correlation function $G(r) = (\theta-\theta^2)^{-1} < \tilde{n}(0)\tilde{n}(r) >_c$ where $\tilde{n}(r) = n(r) - < n >= n(r) - \theta$ and the brackets denote a circular average as well as an average over runs. Here, the density $n(r) = 1$ at a site r at which there is an adlayer atom, and 0 otherwise. Finally, we also calculated the circularly-averaged structure factor $S(k) =< \frac{1}{L^2} | \sum_{\mathbf{r}} \tilde{n}(\mathbf{r}) e^{i\mathbf{k}\cdot\mathbf{r}} |^2 >_c =< \sum_{\mathbf{r}} < \tilde{n}(\mathbf{0})\tilde{n}(\mathbf{r} > e^{-i\mathbf{k}\cdot\mathbf{r}} >_c$.

In order to study the morphology of the islands, we also calculated the radius of gyration of the clusters R_G and studied the dependence of R_G on the island size s. Assuming that $R_G \sim s^{1/d_f}$, we calculated the fractal dimension d_f of the islands as a function of the ratio $R = D/F$ at several values of the coverage θ. We also determined the percolation threshold (i.e. the coverage θ_p at which a single island first percolated through the lattice) as well as the first-layer coverage at percolation θ_{1p} as a function of $R = D/F$.

3. Scaling Theory

Our analysis focuses on the scaling behavior of the monomer density $N_1 \sim R^{-\gamma}\theta^{-\nu}$, island-size distribution N_s (where N_s is the density of islands containing s particles), stable island density $N = \sum_{s\geq 2} N_s \sim R^{-\chi}\theta^{1-z}$, and average island size $S \sim R^{\chi}\theta^{z}$ as functions of the coverage θ and the ratio $R = D/F$ of diffusion rate to deposition rate. The fundamental quantity in the kinetic description of island growth is the island size distribution function $N_s(t)$, which gives the density of islands of size s (where s is the number of atoms or particles in the island) at time t. Since the coverage $\theta = Ft$ is often more convenient for comparison with experiments, we express the time dependence of the island size distribution and its moments in terms of θ. Defining the total island density N (excluding monomers) and the coverage θ by, $N = \sum_{s\geq 2} N_s(\theta)$ and $\theta = \sum_{s\geq 1} sN_s$, then the average island size S can be written in terms of the zeroth and first moments as,

$$S = \frac{\sum_{s\geq 2} sN_s(\theta)}{\sum_{s\geq 2} N_s(\theta)} = \frac{(\theta - N_1)}{N} \quad , \tag{1}$$

where N_1 is the monomer density.

According to the dynamic scaling assumption [6] there exists only one characteristic size in the problem which is the mean island size $S(\theta)$ defined in (1). This implies that one may write the general scaling form,

$$N_s(\theta) = \theta\, S^{-2} f(s/S), \quad (s \geq 2) \quad , \tag{2}$$

where the scaling function $f(u)$ satisfies $\int f(u)\, du = \int f(u) u\, du = 1$ [7]. Assuming that at late time (and for large R), the average cluster size S scales as $S \sim R^{\chi}\theta^{z}$ where $R = D/F$ is the ratio of the diffusion rate D to the deposition rate F. Eq. (2) may then be rewritten in the form,

$$N_s(\theta) = R^{-2\chi}\theta^{1-2z} g_1(\frac{s}{R^{\chi}\theta^{z}}) \quad , \tag{3}$$

where $g_1(u) = \frac{1}{c^2}\, f(u/c)$ where $c = S/(R^{\chi}\theta^{z})$. Alternatively, one may rewrite this in the form,

$$N_s(\theta) = s^{-\tau}\ R^{-\chi/z}\ g_2(\frac{s}{R^{\chi}\theta^{z}}) \quad , \tag{4}$$

where $\tau = 2 - 1/z$ and $g_2(u) = u^{\tau} g_1(u)$. At late time, one expects the walker density N_1 to be much smaller than the coverage θ so that in the scaling regime $N_1 << \theta$ and $S \simeq \theta/N$. This implies the scaling relation for the total island density, $N \simeq \theta/S \sim R^{-\chi}\theta^{1-z}$.

4. Results

Figure 1 shows an overview of our results for the variation of the island density N and monomer density N_1 as a function of coverage θ for a given value of $R = D/F$. As can be seen from Fig. 1, we find that the dynamical behavior of both N and N_1 is divided into four distinct regimes: a low-coverage nucleation regime (L), an intermediate-coverage regime (I), an aggregation regime (A), and a coalescence and percolation regime (C). In the low-coverage nucleation regime, the monomer density is much larger than the island density which is

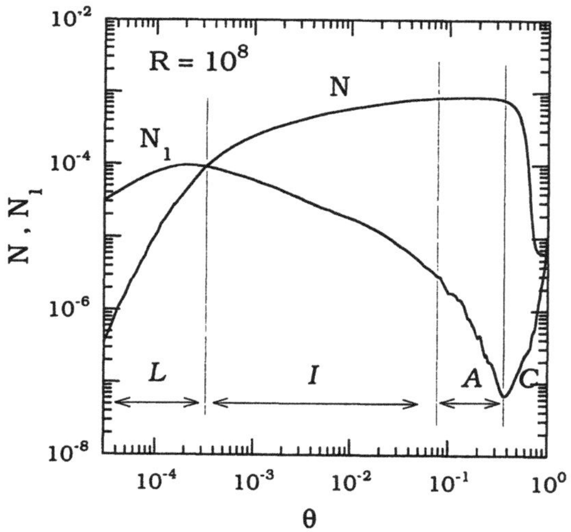

Fig. 1: Log-log plot of unscaled island density N and monomer density N_1 as a function of coverage θ for $R = 10^8$ showing the 4 scaling regimes.

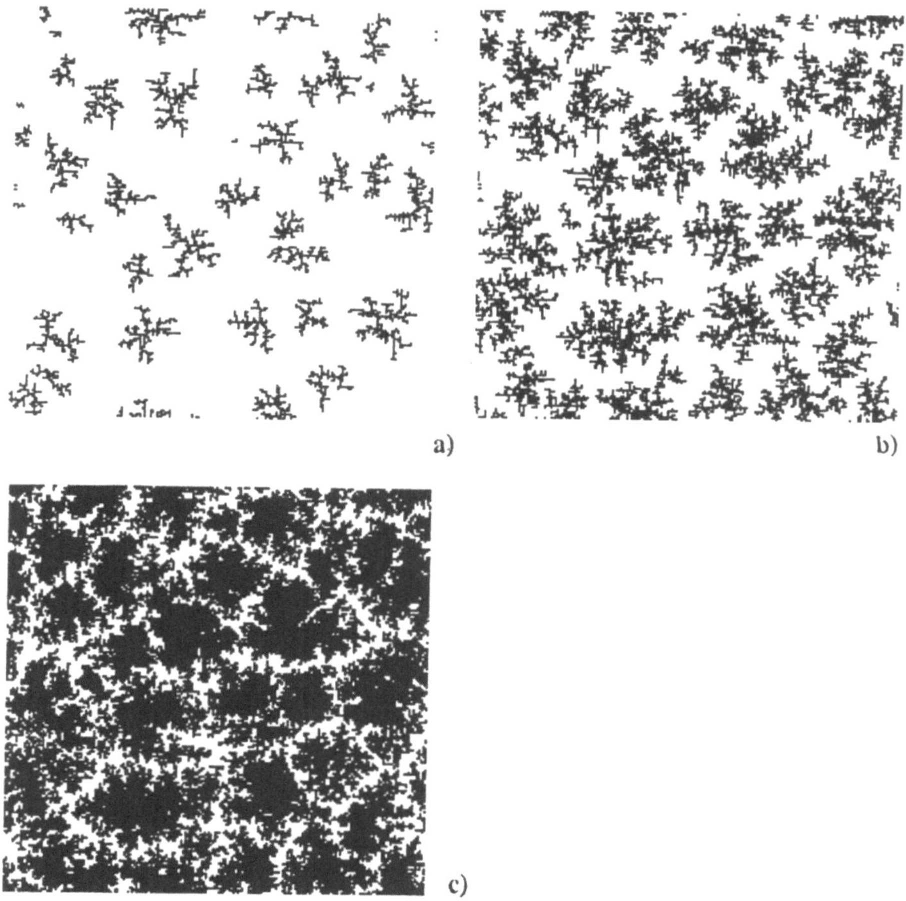

Fig. 2: Adlayer configurations on a 200 × 200 lattice for $R = D/F = 10^8$ and $\theta =$ a) 0.1 b) 0.3 and c) 0.7 ML.

increasing due to the nucleation of new islands. In the intermediate-coverage regime the island density is much larger than the monomer density so that the monomer density is decreasing while the island density continues to increase. In the aggregation regime, the island density and size have become sufficiently large that every adatom deposited lands either near or on top of an island. As a result the island density remains essentially constant while the monomer density is sharply decreasing. Finally, in the coalescence regime, the islands begin to join together and percolate so that eventually second-layer growth occurs and the monomer density again increases.

Figure 2 shows pictures which give an overview of the typical island morphology and distribution for large D/F ($D/F = 10^8$) for three different values of the coverage ($\theta = 0.1, 0.3$, and 0.7) corresponding to the very beginning of the aggregation regime ($\theta = 0.1$), the middle of the aggregation regime ($\theta = 0.3$), and the coalescence and percolation regime ($\theta = 0.7$). As can be seen from the pictures, for relatively low coverage the islands are fractal as is observed in a number of experiments [6,19-22], while at higher coverage they become compact. We now discuss each of these scaling regimes in more detail.

Low-coverage Nucleation Regime $(\theta << R^{-1/2})$

In the low-coverage nucleation regime $\theta << R^{-1/2}$ ($\hat{\theta} << 1$), the island density is much smaller than the monomer density and the island size is very small, so the point-island rate equations [9-11] are expected to hold, except for the inclusion of logarithmic corrections [10] due to diffusion in two-dimensions. Good agreement is found with the predictions, $\hat{N}_1 \simeq \hat{\theta}$, and $\hat{N} \simeq (\hat{\theta}/[ln(2R\theta)])^3$, (where $\hat{N}_1 = R^{1/2}N_1$, $\hat{\theta} = R^{1/2}\theta$).

Intermediate-coverage Regime $(R^{-1/2} << \theta << \theta_I \simeq 0.02)$

For $\theta >> R^{-1/2}$ ($\hat{\theta} >> 1$) , the point-island rate equations [9-11] predict $N_1 \sim R^{-2/3}\theta^{-1/3}$, $N \sim R^{-1/3}\theta^{1/3}$ which implies that the average cluster size $S \simeq \theta/N >> 1$. Consequently, the point-island rate equations must be replaced by rate equations which take into account the island size [12]. We assume that the rate of capture of monomers by an island of size s goes as s^p. For $p = 1/2$ one obtains, $N_1 \sim R^{-3/4}\theta^{-1/2}$ and $N \sim R^{-1/2}log(\theta)$ while for $p = 1/d_f = 0.58$ ($d_f = 1.7$) one obtains $N_1 \sim R^{-0.79}\theta^{-0.58}$, and $N \sim R^{-1/2}(1 - aR^{-q/2}\theta^{-q})$ with $q = 0.16$. Fig. 3 shows relatively good agreement in the intermediate coverage regime with the dynamical behavior predicted for $p = 1/2$ for both N and N_1. However, the R-dependence of the monomer density N_1 agrees better with the point-island prediction $\gamma = 2/3$, while the coefficient of $log(\theta)$ in the fits of the form $N \sim A(R)log(\theta)$ goes as $A(R) \sim R^{-0.41}$.

Aggregation Scaling Regime $\theta_I << \theta < \theta_C$

In the aggregation regime $S \simeq \theta/N$ (since $N_1 << \theta$) so that the ratio d/l of average cluster diameter d to average cluster distance l satisfies $d/l \sim \theta^{1/d_f}$ where d_f is the fractal dimension of a cluster. Consequently, for $\theta > 0.1$ the island size is of the order of the distance between islands while the fraction of the surface occupied by islands is large enough that one must take into account the probability of deposition onto an island as well as onto the substrate. As a result every adatom deposited attaches to an existing island and no new islands are created. This implies the existence of an extended aggregation regime in which the island density $N \sim \theta^{1-z}$ remains approximately constant (see Fig. 1) which implies $z = 1$. In this regime, for large R the island density scales as $N \simeq 0.4R^{-\chi}$ with $\chi \simeq 1/3$ while the monomer density scales as $N_1 \sim R^{-\gamma}$ with $\gamma = 0.7 - 0.8$.

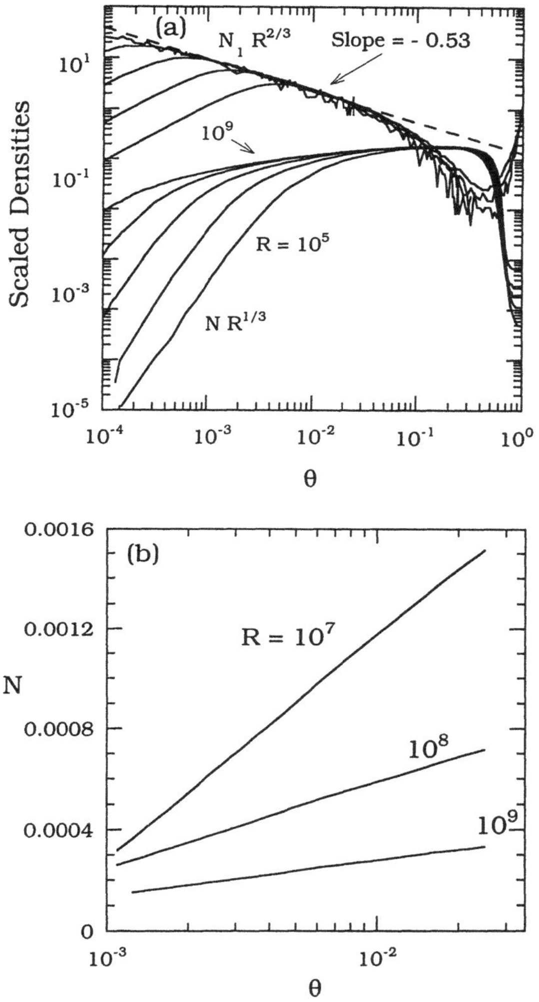

Fig. 3: (a) Variation of the scaled monomer density $N_1 R^{2/3}$ and scaled island density $NR^{1/3}$ over the entire range of coverage θ. Dashed line shows fit to the monomer density in the intermediate-coverage regime. (b) Semilog plot of N vs. θ in the intermediate regime for various R.

Since the island density is constant in this regime, one expects the island size distribution $N_s(\theta)$ to scale with the average cluster size S in agreement with the general scaling form (2), $N_s(\theta) = \theta S^{-2} f(s/S)$. Taking $S \sim R^{\chi}\theta^{z}$, this may be rewritten in the form of Eq. 4. Fig. 4(a) shows scaling plots of this form in the aggregation regime for $\theta = 0.1 - 0.3$ and $R = 10^7 - 10^9$ with $\chi = 1/3$ and $z = 1$. Also shown is the scaled island size distribution for $R = 10^7$

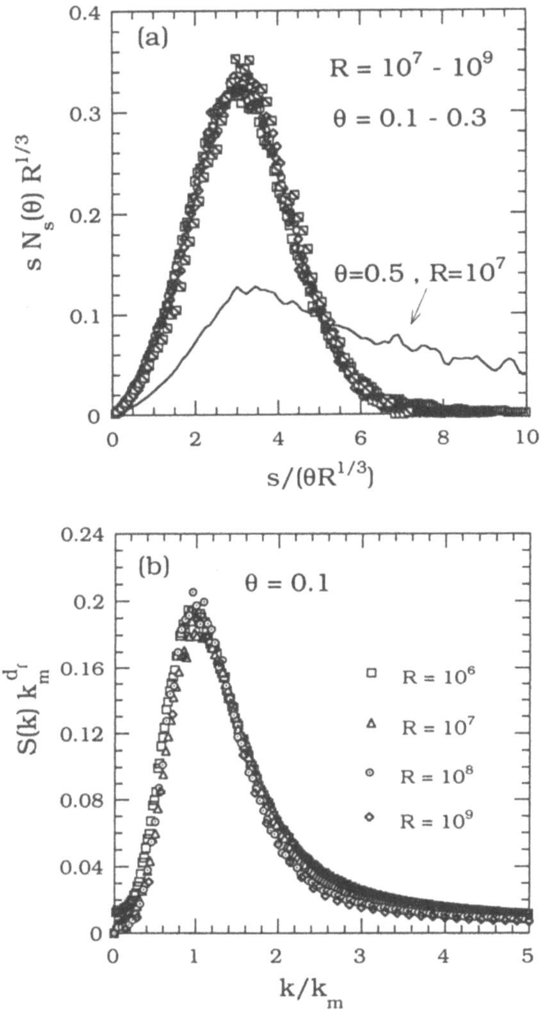

Fig. 4: (a) Scaled island density distribution $N_s(\theta)$ in the aggregation regime (symbols) and in the coalescence regime (solid line) (b) Scaled circularly-averaged structure factor $S(k)$ for $\theta = 0.1$ and $R = 10^6 - 10^9$.

and $\theta = 0.5$ in the coalescence region. In this case the island-size distribution does not scale and there is an extended (approximately power-law) tail due to coalescence.

We have also studied the scaling of the circularly-averaged structure factor $S(k)$ and pair-correlation function $G(r)$ as a function of R and θ. In the aggregation regime, the structure factor $S(k)$ exhibits a peak at $k = k_m(R) \sim$

$N^{1/2}$ corresponding to the average interisland distance. For compact clusters, one expects $G(r)$ to satisfy the scaling form, $G(r; R, \theta) = f_G(rk_m; \theta)$ and $S(k)$ to satisfy, $S(k; R, \theta) = k_m^{-2} f_S(k/k_m; \theta)$ where f_G and f_S are scaling functions. For fractal clusters this may be generalised to [7],

$$S(k; R, \theta) \simeq k_m^{-d_f} f_S(k/k_m)\ f(\theta) \quad , \tag{5a}$$

$$G(r; R, \theta) \simeq k_m^{2-d_f} f_G(rk_m)\ f(\theta) \, , \tag{5b}$$

where d_f is the cluster fractal dimension, $f(\theta) \simeq \theta - \theta^2$, and the R-dependence is given by $k_m \sim R^{-1/6}$. Fig. 4(b) shows the scaling for $S(k)$ for $\theta = 0.1$ and $R = 10^6 - 10^9$, with $d_f = 1.7$. Surprisingly, this scaling form with $d_f = 1.7$ is found to hold for higher coverages (up to $\theta = 0.5$) despite the fact that for $\theta \geq 0.2$ the clusters are relatively compact. This appears to be a consequence of the diffusion-limited process by which the clusters are formed. Similar scaling plots have also been made for $G(r)$ [7].

Coalescence and Percolation Regime

Beyond $\theta_C(R)$ the islands begin to coalesce and eventually percolate, the island density N goes to zero (see Fig. 1(a)) and the monomer density N_1 increases again leading to second-layer growth. We have studied the percolation coverage θ_p as a function of R for $R = 10^1 - 10^9$, using the finite-size scaling form $\theta_p(\infty) - \theta_p(L) = CL^{-1/\nu}$ with $\nu = 4/3$ [13]. System sizes used were $L = 300, 400, 500$, and 1000. Fig. 5 shows that there is non-monotonic behavior

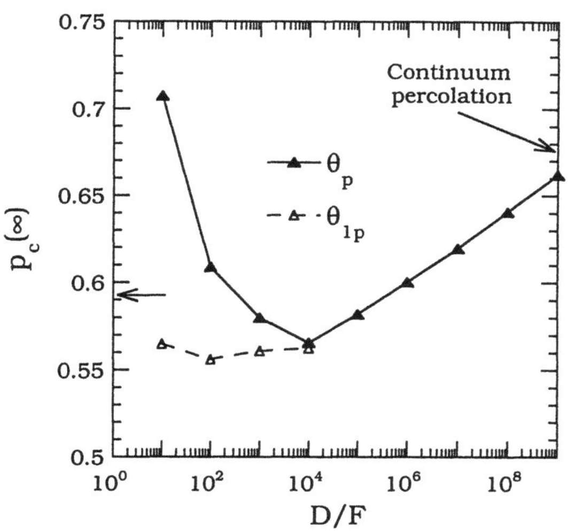

Fig. 5: Percolation coverage θ_p and first-layer coverage θ_{1p} as a function of $R = D/F$ for $R = 10^1 - 10^9$. Arrow at left corresponds to θ_{1p} for $R = 0$.

for θ_p with a minimum near $R = 10^4$. Also shown is the first-layer coverage θ_{1p} at percolation which indicates that there is significant second-layer growth before percolation for $R < 10^4$, but not for $R > 10^5$. This is because for small R the cluster size at percolation is smaller than the diffusion length $l_D \sim R^{1/2}$. For large R, θ_p increases as a function of R due to the increasing compactness (relative smoothness) of the clusters with increasing R. In the limit $R \to \infty$, the clusters will become essentially large smooth circular disks, and assuming no correlations this corresponds to continuum percolation. For $R = 10^9$, $\theta_p \simeq 0.662$ is already close to the continuum percolation limit [14] $\theta_p = 0.676 \pm 0.002$. Thus, we expect that in the limit $R \to \infty$, θ_p may be significantly higher than this value, due to correlation between the clusters. This is similar to what has been seen in thermal evaporation of metal films [15]. There is also a slight non-monotonic behavior in θ_{1p} with a minimum near $R = 10^2$, which is similar to what has been observed [16,17] in models of correlated percolation.

5. Conclusion

We have investigated the evolution, growth and scaling of the island size, density, and distribution as a function of the coverage θ and D/F in a model of molecular beam epitaxy appropriate for the case of dendritic island growth. As expected, we found that the clusters become more compact with increasing coverage, while there is also a slight increase in d_f as R increases. For large R ($R = 10^5 - 10^9$) we found that for $\theta = 0.1$ the islands are fractal ($d_f \simeq 1.7$) while for $\theta = 0.5$ they are compact ($d_f \simeq 2$). Our Monte Carlo results span the range from very low coverage all the way to full-monolayer coverage. Accordingly, we have found four distinct scaling regimes for the scaling of the monomer density N_1 and island density N.

In the early-time low-coverage nucleation regime, we found good agreement with the predictions of the point-island model [10,11]. However, beyond the nucleation regime we found substantial deviations from the dynamical behavior predicted by the point-island model and rate equations. In particular, in the intermediate-coverage regime we found good agreement with the predictions of the generalised rate equations modified to take into account the finite size of islands by including a power-law dependence of the aggregation rate on the island size with $p = 1/2$ [7].

Beyond the intermediate-coverage regime we found, again in contrast to the predictions of the point-island model, an extended aggregation regime for which the island density N remains constant ($z = 1$) while the monomer density N_1 rapidly decreases. We also found excellent scaling [7] of the island-size distribution using the generalised scaling form (2) as well as using Eq. 3 with $z = \tau = 1$ and $\chi = 1/3$ in the aggregation regime for a wide range of R and θ. As expected, the island-size distribution scaling function $f(u)$ ($u = s/S$) is significantly different from that obtained for the point-island model. At higher coverage, however, the island-size distribution no longer scales and develops a long (approximately power-law) tail due to coalescence.

We also investigated the scaling of the circularly-averaged structure $S(k)$ and pair-correlation function $G(r)$ in and beyond the aggregation regime. We found that the scaling of $S(k)$ as a function of R satisfies an anomalous fractal scaling form with $d_f \simeq 1.7$, which we derived for fractal clusters using a simple scaling analysis. This anomalous scaling form with $d_f \simeq 1.7$ also holds for the entire range of coverage in and even somewhat beyond the aggregation regime ($\theta = 0.1 - 0.5$), even though for $\theta \geq 0.2$ the clusters have become significantly more compact. The existence of this anomalous scaling form for $S(\mathbf{k})$ should be useful as an experimental signature for the identification of dendritic islands. For large k, the power-law tail of the structure factor satisfies a different scaling form [7], with an exponent somewhat below the cluster fractal dimension.

Finally, we studied the percolation coverage θ_p as a function of D/F. For small R ($R < 10^4$) there is significant second-layer growth before percolation, and θ_p decreases with increasing R due to the decrease in second-layer growth with increasing R. However, for large R ($R \geq 10^5$) there is negligible second-layer growth before percolation and the percolation coverage θ_p increases with increasing R due to the increasing compactness of the clusters. Thus our results for the percolation behavior as a function of R for large R have important implications for the understanding of multilayer growth. Future work will focus on extending these results to the case of multilayer growth.

Acknowledgements

This work was supported by the National Science Foundation and by the Office of Naval Research.

References

1. J.A. Stroscio, D.T. Pierce, and R.A. Dragoset, Phys. Rev. Lett. **70**, 3615 (1993).
2. E. Kopatzki, S. Gunther, W. Nichtl-Pecher, and R.J. Behm, Surf. Sci. **284**, 154 (1993).
3. Y.W. Mo, J. Kleiner, M.B. Webb, M.G. Lagally, Phys. Rev. Lett. **66**, 1998 (1991)
4. H.J. Ernst, F. Fabre, and J. Lapujoulade, Phys. Rev. B **46**, 1929 (1992).
5. W. Li, G. Vidali, and O. Biham, Phys. Rev. B **48**, 8336 (1993).
6. R.Q. Hwang, J. Schroder, C. Gunther, and R.J. Behm, Phys. Rev. Lett. **67** 3279 (1991); R.Q. Hwang and R.J. Behm, J. Vac. Sci. Technol. B **10**, 256 (1992)

7. J.G. Amar, F. Family, and P.-M. Lam, in *Mechanisms of Thin Film Evolution*, MRS Symposia Proceedings No. 317 (Materials Research Society, Pittsburgh, 1994); Phys. Rev. B (to be published).
8. D. Witten and L.M. Sander, Phys. Rev. Lett. **47**, 1400 (1981).
9. S. Stoyanov and D. Kashchiev, Curr. Topics Mat. Sci. **7**, 69 (1981).
10. L.-H. Tang, J. de Physique I **3**, 935 (1993).
11. M.C. Bartelt and J.W. Evans, Phys. Rev. B **46**, 12675 (1992).
12. J.A. Blackman and A. Wilding, Europh. Lett. **16**, 115 (1991).
13. D. Stauffer and A. Aharony, *Introduction to Percolation Theory* (Taylor and Francis, London, 1992).
14. E. Gawlinski and H.E. Stanley, J. Phys. A **14**, L291 (1981).
15. X. Yu, P.M. Duxbury, G. Jeffers, and M.A. Dubson, Phys. Rev. B **44**, 13163 (1991).
16. S. R. Anderson and F. Family, Phys. Rev. A **38**, 4198 (1988).
17. D.E. Sanders and J.W. Evans, Phys. Rev. A **38**, 4186 (1988).
18. R.L. Schwoebel, J. Appl. Phys. **40**, 614 (1969); G. Ehrlich and F. Hudda, J. Chem. Phys. **44**, 1039 (1966).
19. A. Brodde, G. Wilhelmi, D. Badt, H. Wengelnik, and H. Neddermeyer, J. Vac. Sci. Technol. **B9**, 920 (1991).
20. D.D. Chambliss and R.J. Wilson, J. Vac. Sci. Technol. **B9**, 928 (1991).
21. M. Bott, Th. Michely, and G. Comsa, Phys. Rev. Lett. **70**, 3943 (1993).
22. Th. Michely, M. Hohage, M. Bott, and G. Comsa, Phys. Rev. Lett. **70**, 3943 (1993).

Computer Simulation of the Structure Factor

P.A. Lindgård

Physics Department, Risø National Laboratory, DK-4000 Roskilde, Denmark

Abstract. An introduction is given to the problem of calculating the structure factor by means of numerical simulations. The physical importance of the structure factor is emphasized. Several examples including both lattice gas models, continuous models and exact results are discussed.

1. Introduction

The structure factor is the Fourier transform [1] of the pair correlation function. The latter can easily be obtained in simulations, and there exist fast Fourier transform routines. Is there then any reason to specially discuss the subject of the structure factor? In fact there is, and the following will only be a short introduction to the problem with a number of key formulas and references for further reading.

In recent years the interest of condensed matter physics has shifted towards more and more complex phenomena, disordered structures, non-equilibrium systems and phenomena on a mesoscopic scale. Experimentally, the reciprocal space (q-space) methods as neutron scattering and X-ray scattering (with the high flux synchrotron sources) are pushing for higher and higher q-resolution which in turn means insight into a scale of phenomena between the microscopic (nano-meter) and the macroscopic (μ-meter) scales. These scales have hitherto not been accessible for studies of bulk properties by neither q-space nor real space (r-space) methods such as the various microscopy techniques. The *diffuse* neutron or X-ray scattering, sometimes also called the *small angle* scattering, gives directly the diffuse part of the structure factor $\mathcal{S}_{dif}(q)$. If the scattering is furthermore analyzed in energy transfer one obtains the dynamic structure factor $\mathcal{S}(q,\omega)$, which will only be briefly mentioned here. $\mathcal{S}_{dif}(q)$ is however difficult to interpret and it is often interesting to obtain information on the corresponding r-space behavior. It is here the computer simulation comes in as an important tool, since it can bridge this gap in our understanding. Unfortunately it is not an easy task.

The structure factor is also interesting from another point of view. Computer simulation snapshots or microscopy pictures are not very useful in a statistical sense and one needs to reduce the information to correlation functions. Here the pair correlation function is particularly useful and many theories for this have been developed. In fact, the simulations allow also an investigation of higher order correlation functions, which so far can not be easily measured by any experimental technique. Computer simulation of $\mathcal{S}_{dif}(q)$ is also important in the so called inverse Monte Carlo (MC) methods [2, 3], where the interaction parameters in an effective Ising model are obtained by fitting to experimental data on $\mathcal{S}_{dif}(q)$. The simulations for the Ising model then allow an approximate calculation of the phase diagrams and thermodynamics for an alloy or for another complex system. The theory of $\mathcal{S}_{dif}(q)$, $\mathcal{S}(q,\omega)$ and the pair correlation functions were already developed many years ago for the purpose of analyzing scattering data. This

Springer Proceedings in Physics, Vol. 78
Computer Simulation Studies in Condensed-Matter Physics VII
Editors: D.P. Landau · K.K. Mon · H.-B. Schüttler

literature may not be well known in the computer physics community. It can be found for neutron scattering in [4, 5] and for X-rays in [6], and a particular discussion of $\mathcal{S}_{dif}(q)$ in [7]; a more recent general discussion can be found in [8].

In this paper a number of selected examples will be discussed in which computer simulation has been very important for the understanding of the structure observed by scattering techniques. This holds for the observed [9] imperfect structure of the high T_c superconducting materials as $YBa_2Cu_3O_{6+x}$ where the structural properties seem intimately connected with the superconductivity [10]. The problem was further studied by Fiig *et al* [11]. It also applies [12] to the observed intricate epitaxial structures of adsorbed particles on surfaces [13].

2. The Structure Factor

The *dynamic* structure factor (ω-dependent) and the *static* structure factor (t-independent or ω integrated) are in the Born approximation defined as [4, 5]

$$\mathcal{S}(\mathbf{q},\omega) = \frac{1}{2\pi\hbar N}\int_{-\infty}^{\infty} dt e^{-i\omega t}\langle\hat{\rho}_{\mathbf{q}}\hat{\rho}_{-\mathbf{q}}(t)\rangle, \tag{1}$$

$$\mathcal{S}(\mathbf{q}) = \frac{1}{N}\langle\hat{\rho}_{\mathbf{q}}\hat{\rho}_{-\mathbf{q}}\rangle \overset{trans.sym}{=} \frac{1}{N}\langle|\hat{\rho}_{\mathbf{q}}|^2\rangle, \tag{2}$$

$$\mathcal{S}_{dif}(\mathbf{q}) = \frac{1}{N}\langle\delta\hat{\rho}_{\mathbf{q}}\delta\hat{\rho}_{-\mathbf{q}}\rangle, \qquad \delta\hat{\rho}_{\mathbf{q}} = \hat{\rho}_{\mathbf{q}} - \langle\hat{\rho}_{\mathbf{q}}\rangle, \tag{3}$$

where N is the total number of particles or spins, $\mathbf{q}$ is the wave vector and $\hbar\omega$ the resonance energy for scattering processes which happen at different places in space and time t given by the pair correlation function $\langle\cdots\rangle$. The brackets stand for thermodynamic average values at a given temperature T. The operators entering there are the Fourier transform of occupation variables, or of spin variables for magnetic problems. If there is *translational symmetry* $\hat{\rho}_{\mathbf{q}} = \hat{\rho}^{\dagger}_{-\mathbf{q}}$ the very useful second relation in (2) holds, which reduces the problem to the calculation of a single Fourier sum and not a double sum. The operators are

$$\hat{\rho}_{\mathbf{q}} \overset{latt.}{=} \sum_j^N \int d\mathbf{r}\delta(\mathbf{r}-\mathbf{r}_j)e^{i\mathbf{q}\cdot\mathbf{r}_j} \overset{latt.gas}{=} \sum_j^{N'} P_j e^{i\mathbf{q}\cdot\mathbf{R}_j}, \quad \hat{\rho}_{\mathbf{q}} \overset{magn.}{=} \sum_j^{N'} P_j\mathbf{S}_j e^{i\mathbf{q}\cdot\mathbf{R}_j}, \tag{4}$$

where the first relation is for an arbitrary (continuous) distribution of scatterers, $\mathbf{r}_j$ is the position of a particle. The other relations are for lattice gas models with N' fixed positions $\mathbf{R}_i$, but an occupation probability $P_j = 0$ or 1, where $\sum_{j=1}^{N'} P_j = N$. The spin variable may be a vector or just an Ising or Potts variable with two or several states. When the set of $\mathbf{R}_j$ constitutes a regular lattice and $\langle\hat{\rho}_{\mathbf{q}}\rangle \neq 0$ for $\mathbf{q} = \mathbf{Q}$ there will be Bragg peaks, δ-functions, at the corresponding reciprocal lattice points, $\mathbf{Q}$. If we disregard that part from (2) the remaining part (3) is the *diffuse* scattering $\mathcal{S}_{dif}(q)$, which is the main concern here. For further reading see *e.g.* [4].

In computer simulations one can only treat a cell with a finite N and one often uses *periodic boundary conditions* (PBC) in order to mimic the thermodynamic limit ($N \to \infty$). The structure factor is then not dependent of the cell shape, but is very sensitive to the number of particles N_α considered in a particular dimension. The physical meaningful q-values are discrete and their number is also N_α. There is therefore no substitute for the size of the cell if one wants to consider fine details in $\mathcal{S}_{dif}(q)$, *i.e.* the q-resolution cannot

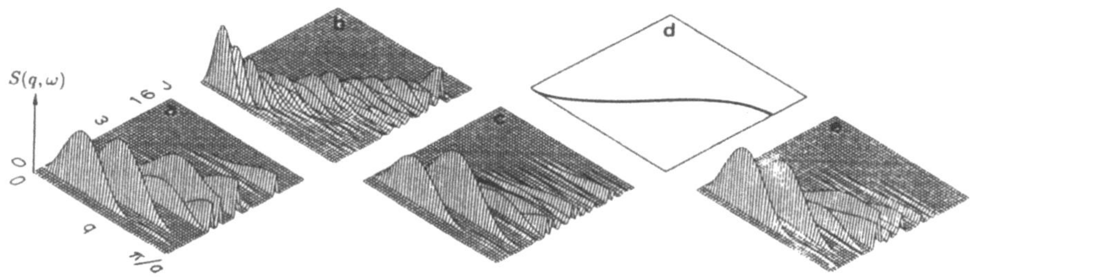

Fig. 1. Dynamical structure factor $\mathcal{S}(\mathbf{q},\omega)$ for $\mathbf{q}$ along the [111] direction, for a number of clusters of different sizes and structures: (a) a *fcc* cluster containing 55 spins, (b) a *fcc* cluster containing 683 spins, and (c) a disordered cluster containing 55 spins. The disturbances in (c) are made by random displacements of the spins by up to 10% of the 55 spin *fcc* lattice with the effect of lifting the degeneracies. In (d) the bulk ($N=\infty$) case is shown. In (e) 10% holes in the *fcc* lattice have been introduced.

be improved by finite size scaling methods (unless one for example wants to study only a smooth $q \to 0$ behavior). The above points can easily be proved by considering the cell as a unit cell in a perfect super structure [14]. Only the Bragg points for the super structure at $q_\alpha = 2p\pi/L_\alpha$, $p = 1, 2, \cdots$, contribute to $\mathcal{S}_{dif}(q)$; $L_\alpha = N_\alpha a$ is the length of the cell in the α-direction, a is the lattice constant, $\alpha = x, y, z$. Since the unit cell can be defined arbitrarily by symmetry allowed operations, the result does not depend on the shape. The intermediate q-values may depend on the shape and must be omitted. With the super cell picture in mind (for example of dimensions $N_x \times N_y \times N_z$) it is easy to convince oneself that one does not get higher resolution in q-space, whether one uses a lattice gas model, where the particles occupy fixed positions in the cell or the continuous model, where the particles occupy arbitrary positions. The relevant q-values are in both cases given by $q_\alpha = 2p\pi/L_\alpha$ although $\mathcal{S}_{dif}(q)$ at these q-values will be different, reflecting the more detailed information in r-space for the continuous case. It is important to note the rule that the detailed r-space information (small r) is found in the q-space at large q-values, and *vice versa*. Therefore, $\mathcal{S}_{dif}(q)$ is periodic in the lattice gas case. The more detailed information in the continuous case is predominantly appearing at large q-values and thus in the absence of periodicity of $\mathcal{S}_{dif}(q)$. We shall only consider (PBC), except for the following example.

If the simulations are done for a finite cluster with for example *open boundary conditions* $\mathcal{S}_{dif}(q)$ must be described by a continuum of q-values. The reason is that all q-values are needed to produce the sharp cut-off and the vanishing values outside the cluster. As an illustration is shown on Fig.1 the free spin excitation spectrum $\mathcal{S}(q,\omega)$ [15] for 'spherical' *fcc* clusters with or without disorder, containing 55 and ~ 700 interacting spins. For an infinite system $\mathcal{S}(q,\omega)$ is a δ-response in both q and ω (*i.e.* at a given $\omega = \omega_k$, only $q = k$ is allowed), while for the finite cluster $\mathcal{S}(q,\omega)$ is continuous in the q-values, for the discrete ω-values corresponding to the 55 possible energy eigenvalues, many of which are degenerate. The continuous functions of q give information both on the spin distribution in the eigenstate as well as on the shape of the cluster.

We shall not discuss $\mathcal{S}(q,\omega)$ further but return to $\mathcal{S}_{dif}(q)$. In higher dimensions than $d = 1$ the Fourier transformation (4) is a very wasteful treatment of the collected data, since only the total projection of the information on the chosen direction of the $\mathbf{q}$ vector is used. Thus, of the N data points only a sequence of $N_\alpha \sim N^{1/d}$ values are Fourier transformed. (On the highly parallel *Connection Machine* it is equally fast to Fourier transform the total system at once and there is not this problem [16]). Suppose one has a $d = 2$ dimensional r-space ensemble for a system which has full symmetry between the x- and y-direction. Then the information is not fully used if one only calculates $\mathcal{S}_{dif}(q = q_\alpha)$, where $\alpha = x$ for example. The correct and optimal result is obtained from the average over the results using both $\alpha = x$ and y. This can of course be generalized to other symmetries, which are given by the interaction Hamiltonian [16]. In order to obtain a good determination of the line shape for $\mathcal{S}_{dif}(q = q_\alpha)$ it is necessary to average over a large number of different ensemble members and fully use the allowed symmetries.

In many cases one is not interested in the actual line shape but just in the characteristic length which is represented by the width of $\mathcal{S}_{dif}(q = q_\alpha)$. A well known, more economic way to determine this approximately is to calculate the moments

$$\langle q^n \rangle = \sum_{q\in\Omega} |q|^n \mathcal{S}_{dif}(q) / \sum_{q\in\Omega} \mathcal{S}_{dif}(q), \tag{5}$$

where Ω is a suitable region around the maximum, usually near a Bragg point. The use of the absolute value allows a determination of the odd moments, which are otherwise usually zero. For example $\langle q \rangle$ and $\langle q^2 \rangle$ are less sensitive to cut-off effects from the choice of Ω than the higher moments. Since q has the dimension of an inverse length $1/L$ this method will provide a measure for L in terms of $\langle q^n \rangle^{1/n}$. The method is of course only useful in a scaling regime where there is only one characteristic length. If one has a theoretical prediction for the line shape (for example a Lorentzian) the method can be used to estimate the correlation length [17].

3. Correlation and Distribution functions

An important property of the pair correlation function is that it is related to the corresponding response function $\chi_\rho(\mathbf{q})$ by the fluctuation dissipation theorem [5]

$$\frac{1}{N}\langle \delta\hat{\rho}_{\mathbf{q}} \delta\hat{\rho}_{-\mathbf{q}} \rangle = k_B T \chi_\rho(\mathbf{q}) \quad = \quad \mathcal{S}_{dif}(\mathbf{q}), \tag{6}$$

where T is the temperature and k_B the Boltzmann constant. The susceptibility function obeys a Dysons equation $\chi(\mathbf{q}) = \mathring{\chi}(T)[1 + \mathcal{J}(\mathbf{q})\chi(\mathbf{q})]$, relating the interacting and the non-interacting susceptibility $\mathring{\chi}(T)$, where $\mathcal{J}(\mathbf{q})$ (to first order) is the Fourier transformed pair interaction function. We assume this can be expanded around a maximum at Bragg point $\mathbf{Q}$, for simplicity we consider here $\mathbf{Q} = 0$. In the mean field approximation $\chi(\mathbf{q})$ can then be written

$$\chi(\mathbf{q}) = \frac{1}{[1/\mathring{\chi}(T) - \mathcal{J}(0)] + [\mathcal{J}(0) - \mathcal{J}(\mathbf{q})]} \overset{q\to 0}{\to} \frac{1}{\mathcal{J}(0)a^2} \frac{1}{\kappa^2 + [\alpha q_x^2 + \beta q_y^2 + \gamma q_z^2]}. \tag{7}$$

Substituting this into (6), (3) shows that the line shape for $\mathcal{S}_{dif}(q)$ near a Bragg point (here 0) is expected to be a Lorentzian with possibly different half-width-half-maxima κ/ϕ with $\phi = \alpha, \beta, \gamma$ reflecting the possible anisotropy of $\mathcal{J}(\mathbf{q})$ in the different q-directions. For the isotropic case ($\alpha = \beta = \gamma = 1$) it is easy to Fourier transform (6), (7) to yield the Ornstein-Zernike correlation function (for $T > T_c$ and $d = 1, 2, 3$)

$$g(r) = \langle \delta\hat{\rho}(0)\delta\hat{\rho}(r) \rangle = \int d^d r \langle \delta\hat{\rho}_{\mathbf{q}} \delta\hat{\rho}_{-\mathbf{q}} \rangle e^{i\mathbf{q}\cdot\mathbf{r}} \propto \frac{e^{-\kappa r}}{r^{\frac{1}{2}(d-1)}} \quad (r \to \infty). \tag{8}$$

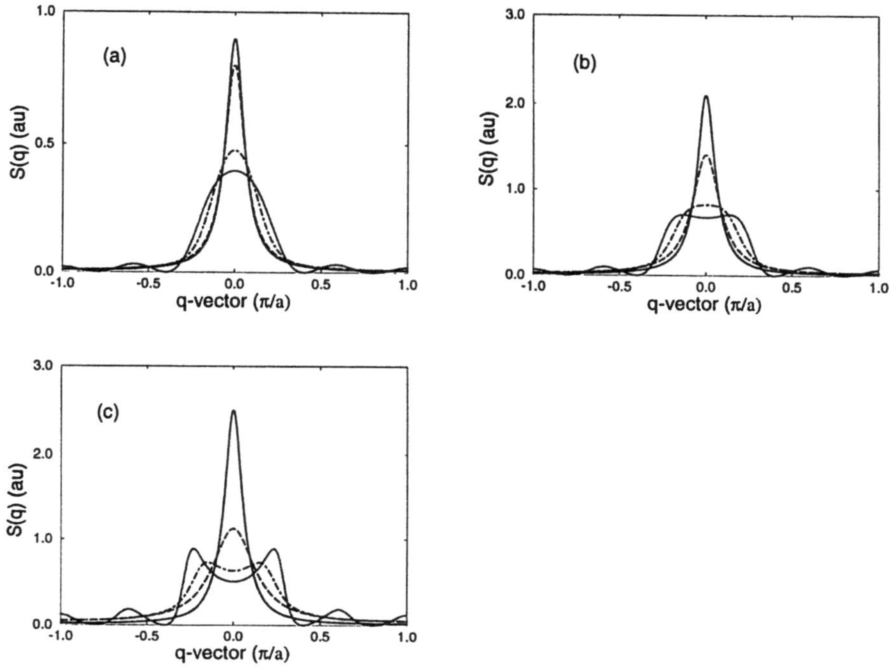

Fig. 2. Calculated, exact line shapes for $\mathcal{S}_{dif}(q)$ for a one-dimensional model including 'excluded volume effect' (EVE) for the various distribution functions $\mathcal{D}(n)$ and for various concentration of particles (a) $x = 0.1$, (b) $x = 0.3$, and (c) $x = 0.5$. The full lines correspond to the Lorentzian line shape (7) neglecting EVE, and also the δ-distribution, including EVE, of equally sized 'hard spheres' of length $\ell = 5$. These are clearly very different and distinguishable. The dashed curve shows the exactly calculated, periodic 'Lorentzian' line shape for the exponential distribution including EVE with $\kappa_{eff} = \kappa/(1-x)$. The dash-dotted line shows the line shape for the Poisson distribution including EVE for the average size $\ell = 5$.

In the critical region $T \sim T_c$ it is a power law of the form $g(r) \propto 1/r^{d-2+\eta}$, where $\eta = 0, \frac{1}{20}$ and $\frac{1}{4}$ for the $d = 4, 3$ and 2- dimensional Ising model [18, 19], and the line shape is not exactly Lorentzian for $d \leq 3$. Experimentally it is common practice to use a Lorentzian square, *i.e.* $\mathcal{S}_{dif}(q) \propto (\kappa^2 + q^2)^{-2}$ for describing the scattering from a disordered domain structure. That corresponds for $d = 3$ to assuming $g(r) = e^{-\kappa r}$.

The relation between $\mathcal{S}(q)$ and the correlation functions follows naturally from the linear response theory. However the modern applications as mentioned above are usually far from the region where that theory is very useful for actually also calculating the correlation functions. It is instructive to calculate the line shape instead from the intuitively simpler distribution function of domain sizes. This can be done exactly in 1-dimension. The problem is to find $\mathcal{S}_{dif}(q)$ corresponding to a distribution of domains (segments filled with particles) of length ℓ with a given distribution $\mathcal{D}(\ell)$ separated by vacancies, and such that no domains overlap (*i.e.* taking into account the 'excluded volume effect'). This was considered by Fratzl *et al* [8] and more recently by Uimin [20] and Fiig *et al* [16]. The exact result can be written in the form

$$\mathcal{S}_{dif}(q) = (1-x)\frac{1-u^2}{1+u^2-2u\cos(\phi)}, \quad ue^{i\phi} = \frac{1}{1-x}\sum_{n=0}^{\infty}\mathcal{D}(n)e^{iq(n+1)a}, \tag{9}$$

where the concentration of particles is x, $0 \leq x \leq 1$. The constraints $\sum_{n=1}^{\infty} n\mathcal{D}(n) = x$ and $\sum_{n=0}^{\infty}\mathcal{D}(n) = 1-x$ are fulfilled when a domain is defined to include a vacancy to the right; then a $n=0$ domain represents a vacancy. Using this Fiig *et al* [16] calculated the line shapes shown on Fig. 2. for the exponential distribution which is expected for $d=1$ according to (8).

Properly normalized, the exponential distribution function can be written $\mathcal{D}(n) = \mathcal{V}\delta_{n,0} + \mathcal{N}e^{-\kappa na}$, where $\mathcal{V} = 1 - xe^{\kappa a}$ and $\mathcal{N} = x(e^{\kappa a}-1)^2/e^{\kappa a}$. This gives indeed a Lorentzian line shape, and most interestingly, this is valid in a very large region of q-space (almost all if the sum from the diffuse scattering centered around all Bragg points is included). This is much more than might be expected from the expansion argument leading to the Ornstein-Zernike form (7). It is therefore a very good approximation to fit to the Lorentzian line shape in order to obtain the correlation length $\xi = 1/\kappa$, at least for $d=1$ and where the physics is suitable. For $x \neq 0$ one finds that the excluded volume effect only results in a new correlations length $\xi(x) = (1-x)/\kappa$, but one finds the same line shape.

Fig. 2. also shows the results for the line shapes for a random distribution of domain sizes with the average length $\ell = 5$, represented by the Poisson distribution $\mathcal{D}(n) = \mathcal{V}\delta_{n,0} + \mathcal{N}\ell^n/n!$ where $\mathcal{V} = 1 - x(1+\frac{1}{\ell})$ and $\mathcal{N} = \frac{x}{\ell}e^{-\ell}$. For small concentration $x \approx 0.1$ an approximate Lorentzian square line shape is found, while for larger $x >$ 0.35 a double-peaked shape is found representing a characteristic separation between the domains related to the average length. The peak separation scales with $1/\ell$ as $C(x)\frac{2\pi}{\ell a}$ with a non trivial prefactor $C(x)$ of order one. A similar result is obtained for a δ-distribution of equally sized domains (hard 'spheres' of length $\ell = 5$) with the distribution $\mathcal{D}(n) = \mathcal{V}\delta_{n,0} + \mathcal{N}\delta_{n,\ell}$, with $\mathcal{V} = 1 - x(1+\frac{1}{\ell})$ and $\mathcal{N} = \frac{x}{\ell}$. Again the peak separation scales with $C'(x)\frac{2\pi}{\ell a}$, where $C'(x) \gtrsim C(x) \sim 1$. For small x the line shape reduces to the well known $\mathcal{S}_{dif}(q) = x(1-x)sin^2(q\ell a)/sin^2(qa)$ function and the zero values are given by $q = \frac{p\pi}{\ell a}$, $p = 2, 3\cdots$. The last examples show that if there is a characteristic domain length in the problem, more sharply defined than the exponential correlation length given by (8), the line shape is not expected to be even close to a Lorentzian.

4. Lattice Gas Model for $YBa_2Cu_3O_{6+x}$

As an example of a lattice gas model study we shall describe the calculation of $\mathcal{S}_{dif}(q)$ for the high temperature super conducting material $YBa_2Cu_3O_{6+x}$ using the ASYNNNI model, which was first proposed for $d=2$ by de Fontaine *et al* [21]. It describes the ordering of oxygen atoms in the CuO_x basal plane of $YBa_2Cu_3O_{6+x}$. It is a lattice gas model given by the Hamiltonian

$$\mathcal{H} = -V_1\sum_{\langle ij\rangle}^{NN} P_iP_j - V_2\sum_{\langle ij\rangle}^{NNNCu} P_iP_j - V_3\sum_{\langle ij\rangle}^{NNNV} P_iP_j - \mu\sum_i P_i, \tag{10}$$

where the sum runs over all oxygen pairs $\langle ij\rangle$, which are nearest neighbors NN or the next nearest neighbors $NNNCu$ across a Cu atom or $NNNV$ across a vacant Cu site. The parameters used are $(V_1, V_2, V_3) = V_0(-1, 0.36, -0.12)$, where $V_0/k_B = 5430K$. The chemical potential μ controls the oxygen content $x = \frac{2}{N'}\langle\sum_i P_i\rangle$, where N' is the available number of oxygen sites. The model has been studied by Monte Carlo simulation for a 200×200 system [11]. On Fig. 3. is shown the calculated $\mathcal{S}_{dif}(q)$ for $\mathbf{q} = (h,k)$ for $x = 0.4$

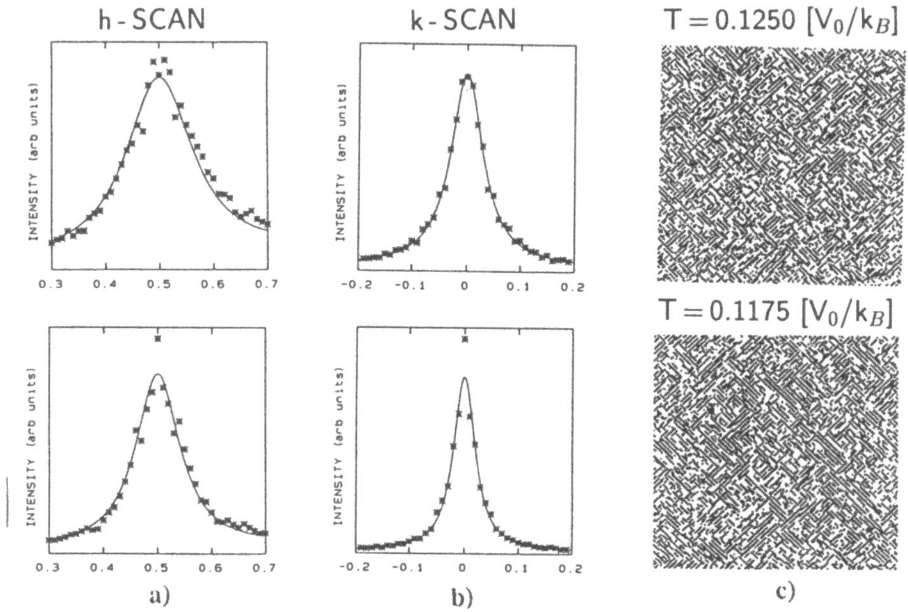

Fig. 3. $\mathcal{S}_{dif}(\mathbf{q})$ for $YBa_2Cu_3O_{6+x}$ for $x = 0.4$. In (a) and (b) are shown the simulated (×) and Lorentzian fitted (line) results for $\mathbf{q} = (\frac{1}{2}, 0)$ in the h-, and k-directions, corresponding to the a- and b-lattice directions which are diagonal in (c), where (c) shows the corresponding snapshots of the oxygen configuration for a 200 × 200 system, updated by Kawasaki dynamics.

and a temperature $T > T_c = 0.1052V_0/k_B$ and one near T_c. The data, which are fitted to a sum over Lorentzians from all Bragg peak positions, show an anisotropy in h and k reflecting the anisotropy in the model. It is clear that in this Ising type problem, $\mathcal{S}_{dif}(q)$, for $T \gtrsim T_c$, the line shape is well described by the Lorentzian form (7). The line shapes and the line widths are furthermore in very good agreement with those experimentally observed [9]. To the left is shown the corresponding r-space snapshots, where the black dots show the oxygen position. It demonstrates that the oxygens form domain structures of long parallel chains. In the experiments the structure freezes into a metastable micro domain structure resembling that at $T = 0.1175V_0/k_B$, when cooled to the much lower superconducting temperatures $T < 90K$. In MC studies it is possible to reach the one domain equilibrium structure with all chains parallel (which is the perfect so called ortho-II structure for $x = 0.5$, having Bragg peaks at $\mathbf{q} = (\frac{1}{2}, 0)$ and $(1, 0)$ etc.) and a distribution of vacancies, since at $x = 0.4$ there is a lack of oxygen atoms for the ortho-II structure. The simulations have also been extended to $d = 3$ [22]. Although the inter plane coupling V_4 is very small, $V_4 = 0.02V_0$, the inclusion of it has a drastic effect on the value of T_c (giving a $\sim 30\%$ increase in T_c) because of the dimensional effects on the critical behavior in going from $d = 2$ to $d = 3$ [18, 19].

5. Continuous Monte Carlo Model for Adsorbed Particles on Graphite

Graphite has one of the most smooth microscopical surfaces available in nature. It has therefore been used experimentally for studying ordering phenomena in $d = 2$ of particles which weakly bind to the surface while having a pairwise mutual interaction. To MC simulate this a continuous MC method was developed [14]. However, the surface is not perfectly smooth, but corrugated in a honeycomb pattern with a hexagonal symmetry. The graphite structure interferes with the natural hexagonal structure of the adsorbed particles and gives rise to modulated adsorbate structures, that are deformed and epitaxially rotated relative to the graphite structure. Very accurate experimental data [13] for the rare gas atoms adsorbed on graphite at variable coverage x, showed unexpected and unexplained patterns in $\mathcal{S}_{dif}(q)$. A MC study [12] demonstrated that a hitherto neglected mechanism in which vacancies participate actively in the structure formation is the solution to the problem. We used a system with $N' = 8100$ honeycomb sites, which can be subdivided into three sublattices, A, B and C; on this surface is distributed between $N = 1000 - 4000$ particles with continuously varying positions $\mathbf{r}_j$.

Fig. 4. shows the calculated $\mathcal{S}_{dif}(q)$ together with the corresponding r-space snapshots for various coverage $x = 1.10, 1.27, 1.47$ where $x = 1$ corresponds to a coverage where every third of the graphite honeycomb sites is occupied, for example sublattice A. The corresponding reciprocal lattice is shown as lines in Fig. 4a-c. The corners are at the reciprocal lattice vectors $\mathbf{Q}_{HK}$ with $(H, K) = (0,0)$ at the lower left and $(1,1)$ in the middle for the underlying graphite surface. At low and high coverage d) and f) multi domain, rotated structures are formed, whereas at an intermediate coverage e) a new single domain structure is formed. This is interesting because the structure must be described by two principal ordering q-vectors simultaneously, shown as the bright spots in the lower left corner of b). These spots are almost the same as those seen in c). However, for that structure they correspond to a sum of the contributions from the two rotated

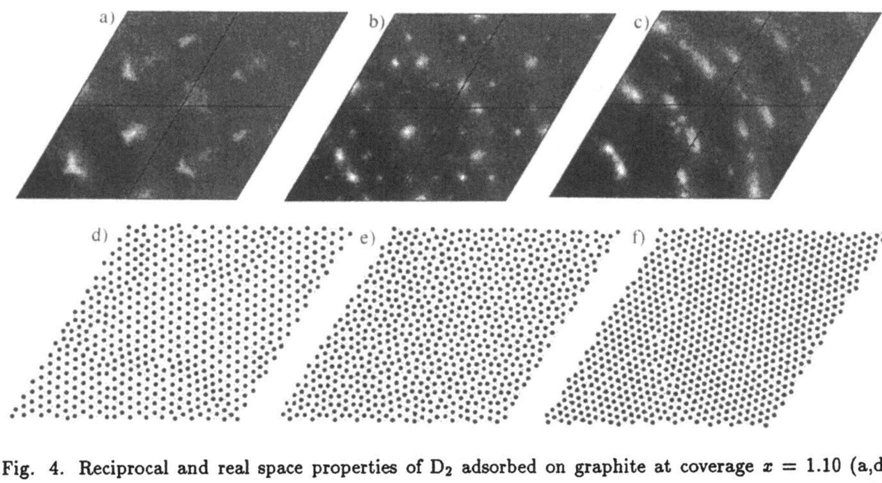

Fig. 4. Reciprocal and real space properties of D_2 adsorbed on graphite at coverage $x = 1.10$ (a,d), $x = 1.27$ (b,e), and $x = 1.47$ (c,f). Top panel, intensity plots of the two-dimensional structure factor $\mathcal{S}(\mathbf{q})$ in a logarithmic scale. The dark region in the lower left corner $(H, K) = (0,0)$ has small intensity $(\rightarrow 0)$, the white regions have the highest intensity. The δ-functions at $(H, K) = (1,0), (1,1)$ etc. are present but not visible in this representation. $(H, K) = (1,1)$ is in the middle of the pictures. Lower panel, a small section of the simulated systems. (• particles, · substracte potential wells).

Fig. 5. Small sections of real space snapshots for D_2 on graphite. (a) For the so called γ-phase and (b) for the 4×4 structure. To the left is shown the continuous representation as used in the MC simulation, and to the right the same structure in the centered representation. (• particles, · graphite potential wells).

domains, seen in f). This is a demonstration of a general problem with the interpretation of experimental $\mathcal{S}_{dif}(q)$, in particular if only a limited region of q-space (small q) can be measured. It is not possible to distinguish between a multi-q (one domain) structure as shown in e) and a superposition of single-q structures (multi domains) as represented by f). In the MC simulations we were able to explore a much larger region in q-space than so far experimentally possible.

Another interesting feature of $\mathcal{S}_{dif}(q)$ is that it is intimately related to the energy function of the model with pair interactions between the particles because of the relation to the pair correlation function (6). For the graphite problem the Hamiltonian is

$$\mathcal{H} = \sum_{\langle ij \rangle} V_{ij} \hat{\rho}(\mathbf{r}_i) \hat{\rho}(\mathbf{r}_j) + \sum_i U(\mathbf{r}_i) \hat{\rho}(\mathbf{r}_i), \tag{11}$$

where V_{ij} is the pair interaction (*e.g.* the isotropic Lenard-Jones interaction) between the adsorbed particles, and $U(\mathbf{r}_i)$ is the corrugation potential of the graphite surface with the honeycomb symmetry, and finally the position operator is $\hat{\rho}(\mathbf{r}_i) = \int d\mathbf{r}\, \delta(\mathbf{r} - \mathbf{r}_i)$, see (4). By Fourier transforming this Hamiltonian the energy can be written

$$\langle \mathcal{H} \rangle = \int \mathcal{V}(q)\, \mathcal{S}(q)\, d\mathbf{q} + \int \sum_{HK} \mathcal{U}_{HK} \delta(\mathbf{q} - \mathbf{Q}_{HK})\, \langle \hat{\rho}(\mathbf{q}) \rangle\, d\mathbf{q}, \tag{12}$$

where $\mathcal{V}(q)$ and $\mathcal{U}_{HK}$ are the Fourier transforms of V_{ij} and of $U(\mathbf{r}_i)$, respectively. The form (12) is very useful for discussing the competition between the interactions V_{ij} and $U(r_i)$ and gives a theoretical understanding of the structures observed both experimentally and in the MC simulation for the molecules D_2 and H_2 on graphite [12].

As discussed in section 2. $\mathcal{S}(q)$ is not obtained with higher q-resolution when using a continuous method, which is much more demanding in computer time. The reason for using this is the physics of the problem, namely the importance of the long range *'elastic forces'* which act between the different domains. In fact there are no such forces in nature, all interactions are described by the (short-ranged) interactions already in the Hamiltonian, for example $\mathcal{H}$ (11). The notion arises only in the lattice gas models such as (10). To see this, consider a row of p adsorbed atoms. If the first is in the center of a honeycomb A, the neighbor will be slightly displaced, its neighbor displaced by the same amount, relatively etc., finally the p'th neighbor is closer to the center of another honeycomb, say of type B. This is easily and exactly described in the continuous model (11). However if every particle is only characterized by which type of cell (A or B) it belongs to, all particles $n \leq p-1$ are of type A, and suddenly the p'th particle is of type B, and a sharp domain wall is created. This corresponds just to the lattice gas model. On Fig. 5. is shown the same configuration for the continuous structure and for the centered structure, where cells containing a particle are simply plotted with it at the center of the cell. $\mathcal{S}_{dif}(q \leq Q_{11})$ for these different realizations is almost the same. However in the lattice gas model using interactions as in (11) there is no interaction between the domain walls and the MC simulation for the centered version will not produce the correct result. It is very complicated to put the missing 'elastic' interaction (which depends on the actual structure) into a lattice gas model as for example the ASYNNNI model(10). Such models can take into account the length and topography (vertices) of domain walls, but neglect the effect of the interactions between them.

6. Phase Separation Problems and Scaling

In the problem of phase separation or spinodal decomposition one usually expects a self-similar behavior of the domain distribution as the system coarsens with increasing time t. The scaling behavior occurs at very large length scales ($L \gg a$), which is not easily accessible by the direct computer modeling. The interest is in very fine details of $\mathcal{S}_{dif}(q)$ for $q \ll \pi/a$ around $q = 0$ or a Bragg point Q. For a random distribution of hard spheres Porod [23] showed the $\mathcal{S}_{dif}(q) \propto q^{d+1}$ for $q \to \infty$, yet smaller than Q of course. This behavior is clearly not obeyed by the Lorentzian (7). Experiments and computer simulations indicate, although not conclusively, that also the small q-limit is unexpected, namely $q \propto q^4$, again in contradiction to the mean field behavior (7) and also to the more complicated line shapes shown in Fig.2. To obtain $\mathcal{S}_{dif}(q \to 0) \to 0$ one needs a local conservation of matter as discussed by Fratzl *et al.* They showed that it is possible to obtain the behavior in a $d = 1$ model with excluded volume, like the ones discussed in section 3. However, one has to distribute both matter and vacancies in a correlated manner, for example for $x = 0.5$ by dividing the chain in Poisson distributed boxes and filling each only half around the center. This reflects the 'snake' like patterns often seen in phase separation problems, see Fig. 6. Fratzl *et al* have suggested the following scaling function for the spherically averaged structure factor $\mathcal{S}_{dif}(q, t)$, for increasing coarsening time t.

$$\mathcal{S}_{dif}(q,t) = \mathcal{S}_{dif}(q_{max}(t),t) \;\; \mathcal{F}(\tilde{q}), \quad \tilde{q} = q/q_{max}(t), \tag{13}$$

where $q_{max}(t)$ is q at the maximum of $\mathcal{S}_{dif}(q,t)$ at a given time t, look for examples in Fig. 2. The q-dependence should therefore be given by a universal function $\mathcal{F}(\tilde{q})$ apart from the t-dependent prefactor. Taking into account the expected limiting behavior they suggested the following empirical form for $d = 3$

$$\mathcal{F}(\tilde{q}) = \frac{a\tilde{q}^4}{\tilde{q}^4 + c} \, \frac{b}{b + (\tilde{q}^2 - D)^2}. \tag{14}$$

There are in fact only two free parameters, b and D (thus for $D = 1: \; a = 1, \; c = 0$). For $d = 2$ the form of $\mathcal{F}(\tilde{q})$ is more complicated. Fratzl *et al* found that their MC data were well obeying the suggested scaling form. In Fig. 6. are shown their MC results for $d = 2$ and $x = 0.5$ for a 512×512 size Ising model. This example shows that the line shape form is indeed much more complicated than the Lorentzian one (7), and as given by the distributions discussed in section 3., see Fig. 2.

Another approach to study the mesoscopic phenomena of phase separation is to use partial differential equations for a phenomenological description of coarse grained systems. The equations can be derived from the Langevin equation

$$\frac{\partial \psi}{\partial t} = -\eta \frac{\partial f}{\partial \psi}, \tag{15}$$

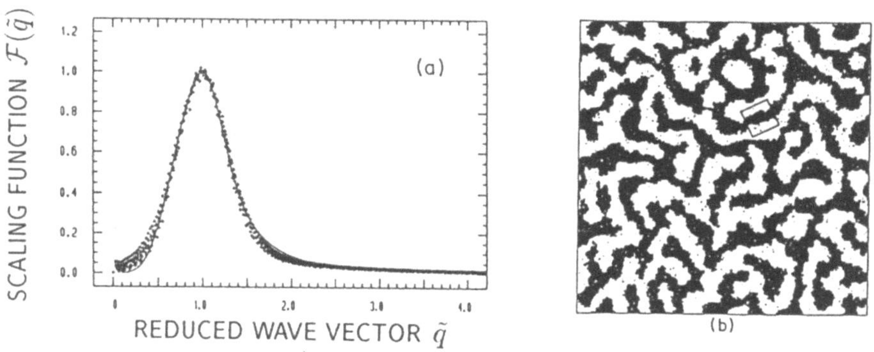

Fig. 6. (a) Scaling function $\mathcal{F}(\tilde{q})$ for the Ising model of phase separation at $x = 0.5$ (system size 512×512, $d = 2$). Different signatures correspond to different times t. (b) real space snapshot showing 'snake' like domains. The box indicates the length corresponding to $2\pi/q_{max}$. After Fratzl *al* [8]

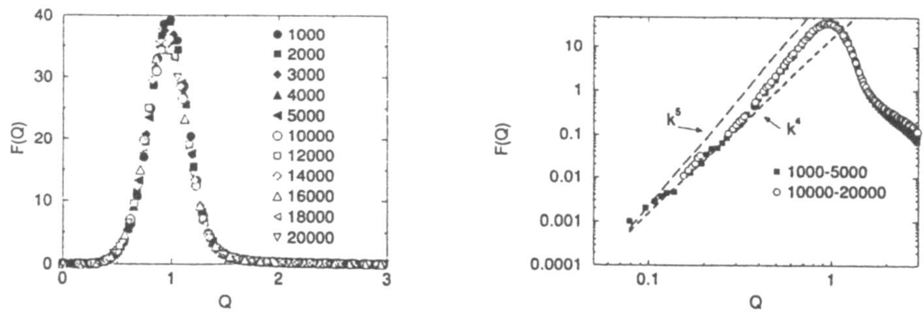

Fig. 7. Left, the scaling function Eq. (13) (with $Q = \tilde{q}$) plotted for various time steps for the binary alloy sinodal decomposition at $x = 0.5$ for a $192 \times 192 \times 192$ 3d-system. Note there is no change in shape from the earliest to the latest time data. To the right, the same data plotted in a double logarithmic scale. This shows that it has been possible to detect interesting deviations from the expected theoretical behavior. After Shinozaki and Oono [25]

where ψ is the order parameter (here the concentration difference), η a kinetic coefficient and f the corse grained free energy. By expanding f near the transition point for small ψ Langer, Bar-on and Miller [24] suggested the following equation for conserved dynamics

$$\frac{\partial \psi}{\partial t} = -\nabla \cdot (1 - a\psi^2)\nabla(D\Delta\psi + \tau\psi - g\psi^3), \tag{16}$$

where τ and g are free energy parameters and D the diffusion constant, and $a \neq 0$ is introduced to assure Kawasaki dynamics. The problem with this approach is that one has lost the direct link to the parameters in the Hamiltonian and that the equation has only a limited region of application; on the other hand it is good for studying the asymptotic and scaling behaviors, which are supposed to be universal. It is solved by discretized space-time models or cell-dynamical systems[25]. A particularly nice application of this in a study of the limiting behavior for $S_{dif}(q,t)$ for spinodal decomposition of alloys and liquids was done recently by Shinozaki and Oono [25]. In their paper they discuss in detail methods to improve the accuracy of a determination $S_{dif}(q,t)$ and emphasize in particular that, although it is not a self-averaging quantity, the best result is obtained using a single large system rather than using the average of several small ensemble members. As an illustration we show in Fig. 7. the calculated scaling function $\mathcal{F}(Q=\tilde{q})$ on a linear and a double logarithmic scale. With the impressive accuracy they have obtained, deviations from the expected theoretical limiting behaviors are noticeable and interesting. This clearly demonstrates that $S_{dif}(q)$ contains a lot of information and that it is difficult to describe theoretically.

7. Advantages and Difficulties in Simulations of the Structure Factor

When $S_{dif}(q)$ is used experimentally, for example to determine domain growth, phase separation kinetics, the coarsening of a spinodal decomposition in alloys or fractal properties of gels, it is very difficult to avoid an influence of unwanted effects from impurities and the like. Here of course the computer models have the big advantage. One knows the Hamiltonian $\mathcal{H}$, and can compare directly with ideal theories. A question is if the Hamiltonian has much to do with the actual problem. For detailed properties, such as surface reconstruction or complicated alloy structures the details in the $\mathcal{H}$ may be important in order to obtain a correct description. A simulation may then be useful for obtaining the corresponding, relevant interaction terms. For more statistical properties and in regimes of scale invariance only the main symmetries of $\mathcal{H}$ probably matter.

In section 4. it was mentioned that it was possible to obtain the equilibrium structures in the MC simulations contrary to what was possible in the experiments. The case is usually the reverse, as experienced in the adsorbed particle case. The reason is that the Monte Carlo 'time' available in nowadays computers can still only follow a short time relative to that available in most experiments. This is particularly a problem if the relaxation time increases exponential below a certain temperature, as happens for example at the glass transition. In these situations computer simulations are very difficult because of the long lived mestastable states. The advantage is on the other hand that one can study the properties of the constructed ideal states which perhaps cannot be produced experimentally.

8. Conclusion

It has only been possible here to discuss a few examples of the calculation of the structure factor, hopefully showing the diversity of problems at hand. Experimentally a lot of effort is put into building large facilities for measuring this important quantity. With the development of the large parallel computers there is a possibility that the simulations can match this effort. Finding other methods than brute force for improving the resolution in reciprocal space in simulations may be a challenge for the future.

Acknowledgments

It is a pleasure to thank my collaborators T. Fiig, P.V. Hendriksen and E.Vives for having carried out almost all the numerical work discussed here. J.S. Pedersen is thanked for discussions concerning small angle scattering.

References

[1] D. C. Champeney, *Fourier Transforms and their Physical Applications* (Academic Press, London, 1973).

[2] W. Schweika and H.-G. Haubold, Phys. Rev. B **37**, 9240 (1988).

[3] L. Reinhard, B. Schönfeld, and G. Kostorz, Phys. Rev. B **44**, 1727 (1990).

[4] G. L. Squires, *Thermal Neutron Scattering* (Cambridge Press, Cambridge, 1978).

[5] S. W. Lovesey, *Theory of Neutron Scattering from Condensed Matter* (Oxford University Press, New York, 1984).

[6] A. Guinier, *X-Ray Diffraction* (W. F. Freeman and Co., San Fransisco, 1963).

[7] M. A. Krivoglaz, *Theory of X-Ray and Thermal-Neutron Scattering by real Crystals* (Plenum Press, New York, 1969).

[8] P. Fratzl, J. L. Lebowitz, O. Penrose, and J. Amar, Phys. Rev. B **44**, 4794 (1991).

[9] T. Zeiske *et al.*, Nature **353**, 542 (1991).

[10] H. F. Poulsen *et al.*, Nature **349**, 595 (1991).

[11] T. Fiig *et al.*, Physica C **217**, 34 (1993).

[12] E. Vives and P. A. Lindgård, Phys. Rev. B **47**, 7431 (1993), and E. Vives and P. A. Lindgård, Surf. Sci. Lett. **284**, L449 (1993).

[13] H. Freimuth, H. Weichert, H. P. Schieldberg, and H. J. Lauter, Phys. Rev. B **42**, 587 (1990).

[14] E. Vives and P. A. Lindgård, Phys. Rev. B **44**, 1318 (1991).

[15] P. V. Hendriksen, S. Linderoth, and P. A. Lindgård, Phys. Rev. B **48**, 7259 (1993).

[16] T. Fiig *et al.*, , to be published.

[17] P. A. Lindgård and O. G. Mouritsen, Phys. Rev. B **41**, 688 (1990).

[18] H. E. Stanley, *Intoduction to Phase Transitions and Critical Phenomena* (Oxford, New York, 1971).

[19] P. A. Lindgård, in *Topics in Current Physics, Vol. 6., Neutron Diffraction*, edited by H. Dachs (Springer Verlag, Berlin, Heidelberg, 1978), Chap. 6, a short review about critical phenomena.

[20] G. Uimin, Phys. Rev. B , to be published.

[21] D. de Fontaine, L. T. Wille, and S. C. Moss, Phys. Rev. B **36**, 5709 (1987).

[22] T. Fiig, N. H. Andersen, P. A. Lindgård, and O. G. Mouritsen, , to be published.

[23] G. Porod, Colloid-Zeitschrift **125**, 51 (1952).

[24] J. S. Langer, M. Bar-On, and H. D. Miller, Phys. Rev. A **11**, 1417 (1975).

[25] A. Shinozaki and Y. Oono, Phys. Rev. E **48**, 2622 (1993).

Part II

Quantum Systems

The Density-Matrix Renormalization Group for Fermion Systems

R.M. Noack[1], *S.R. White*[1], *and D.J. Scalapino*[2]

[1]Department of Physics, University of California, Irvine, CA 92717, USA
[2]Department of Physics, University of California,
Santa Barbara, CA 93106, USA

Abstract. We discuss techniques of the density matrix renormalization group and their application to interacting fermion systems in more than one dimension. We show numerical results for equal–time spin–spin and singlet pair field correlation functions, as well as the spin gap for the Hubbard model on two chains. The system is a gapped spin liquid at half–filling and shows weak algebraic d-wave–like pair field correlations away from half–filling.

1 Introduction

The numerical renormalization group was developed by Wilson [1] and used by him to solve the one impurity Kondo problem. The technique was subsequently applied to a number of quantum lattice systems [2, 3] such as the Hubbard and Heisenberg models, but with little success. A suggestion by Wilson [4] to investigate why the technique fails for the simplest quantum lattice system, the one-dimensional electron gas, led to the development of a number of new techniques to overcome the difficulties of the numerical RG for this simple system [5]. White [6] was able to generalize one of these techniques to interacting systems, applying it successfully to one dimensional quantum spin systems. This technique has come to be known as the density matrix renormalization group (DMRG).

This paper describes our our current efforts to apply the DMRG to fermion systems in more than one dimension, and in particular to the Hubbard model. So far, we have successfully applied the method to the Hubbard model on one and two chains [7]. Here we discuss the details of the methods we have developed for the two–chain Hubbard model and show results for equal–time pair field and spin–spin correlation functions and for the spin gap for the half–filled and doped systems on lattices of up to 2×32 sites.

At half–filling, both pair field and spin–spin correlations decay exponentially, with the spin correlations having a longer correlation length. There is a spin gap present at half filling which gets smaller as the system is doped, but persists down to band fillings of $\langle n \rangle = 0.75$. For the doped system, the largest pair field correlations are ones in which a spin singlet pair is formed on adjacent sites on different chains. The pair field symmetry is d–wave–like in that the pair field wave function has opposite sign along and between the chains. The

Springer Proceedings in Physics, Vol. 78
Computer Simulation Studies in Condensed-Matter Physics VII
Editors: D.P. Landau · K.K. Mon · H.-B. Schüttler

form of the decay of the pair field correlations for the doped system is algebraic with a form close to that of the noninteracting system, which decays as ℓ^{-2}.

2 The Density Matrix Renormalization Group

The goal of the procedures discussed here is to find the properties of the low-lying states of a quantum system on a particular finite lattice. One way to do this would be to diagonalize the Hamiltonian matrix using a sparse matrix diagonalization method such as the Lanczos technique. However, for interacting quantum lattice systems, the number of states grows exponentially with the size of the lattice. Since exact diagonalization techniques must keep track of all the states, the maximum possible lattice sizes for interacting Hamiltonians is severely limited. It is therefore desirable to develop a procedure in which the Hilbert space of the Hamiltonian can be truncated in a controlled way so that only states that are important in making up the low-lying states of the system are included in a diagonalization. The DMRG provides a procedure for building up such a representation of the Hamiltonian matrix, which is then diagonalized to provide the properties of the low-lying states of the finite system.

The strategy of the DMRG is to build up a portion of the system (called the system block) using a real–space blocking procedure and then truncate the basis of its Hamiltonian after each blocking. In this way, the size of the Hilbert space is kept manageable as the system block is built up. The key idea is the method of truncating the Hilbert space of the system block in a controlled way. This is done by forming the reduced density matrix for the system block, given an eigenstate of the entire lattice. Let us first examine this procedure.

2.1 The Density Matrix Projection

Consider a complete system (the "universe"), divided into two parts, the "system", labeled by coordinate i, and the "environment" [8], labeled by coordinate j. If we knew the exact state ψ_{ij} of the universe, (assuming the universe is in a pure state) the prescription for finding the state of the system block would be to form the reduced density matrix of the system as part of the universe,

$$\rho_{ii'} = \sum_j \psi_{ij} \psi^*_{i'j} . \tag{1}$$

The state of the system block is then given by a linear combination of the eigenstates of the density matrix with weight given by the eigenvalues. It is shown in Ref. [6] that the optimal reduced basis set for the system block is given by the eigenstates of the density matrix with the largest weights. The sum of the density matrix weights of the discarded states gives the magnitude of the truncation error.

3 Algorithms

The density matrix projection procedure gives us a way of truncating the basis set of the matrix for the system block in a controlled way as degrees of freedom are added to the system. The projection procedure of the previous section assumes that the wavefunction ψ_{ij} of the system is known. Of course, finding ψ_{ij} is the goal of the DMRG procedure, so effective algorithms must iteratively improve approximations to ψ_{ij}. We will first discuss the algorithms for one-dimensional systems, as developed in Ref. [6].

In order to perform the density matrix projection procedure, we form the Hamiltonian for a "superblock" which is an approximation to the universe of the previous section. In this case, the superblock will describe a one-dimensional lattice of L sites, with, for example, a Heisenberg or Hubbard Hamiltonian. The superblock configuration used for the one–dimensional algorithms developed in Ref. [6] is shown in Fig. 1. The superblock is formed from an approximate Hamiltonian for the system block containing ℓ sites (labeled by B_ℓ), the Hamiltonians for two single sites which can be treated exactly, represented by solid circles, and an approximate Hamiltonian for the rightmost ℓ' sites, labeled by $B^R_{\ell'}$. Thus, the superblock contains $L = \ell + \ell' + 2$ sites. The algorithm proceeds as follows:

1. The superblock Hamiltonian is diagonalized using a Lanczos or similar exact diagonalization technique to find a particular target eigenstate ψ_{ij}.
2. The reduced density matrix is formed for the system block $B'_{\ell+1}$ using Eq. (1).
3. The density matrix is diagonalized using a dense matrix diagonalization.
4. The Hamitonian for $B'_{\ell+1}$ is transformed to a truncated basis formed by the m highest weighted eigenstates of the density matrix.
5. This approximate Hamiltonian, labeled by $B_{\ell+1}$ is used as a starting point for the next iteration, starting with step 1.

Initially we choose ℓ to be small enough (a single site, for example) so that the Hamiltonian for B_ℓ can be treated exactly. The system block then grows by a single site at each iteration, but the dimension of its Hilbert space remains m. A single site only is added to B_ℓ in order to minimize the size of the superblock

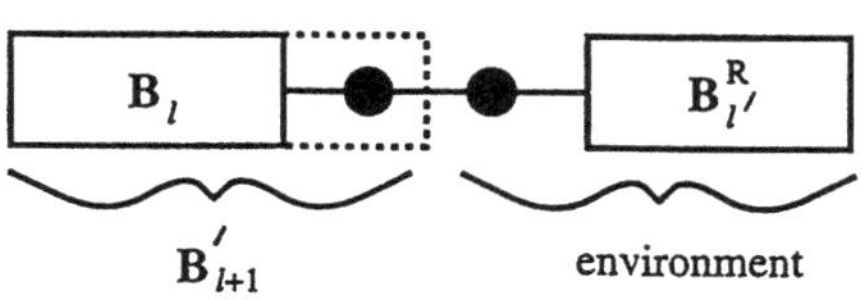

Figure 1: The superblock configuration for the one–dimensional algorithms. The solid circles represent single sites treated exactly and the boxes represent approximate Hamiltonians representing ℓ and ℓ' sites.

Hamiltonian, whose dimension will be n^2mm' where n is the number of states per site, and m' is the size of the basis for $B^R_{\ell'}$.

3.1 The Infinite System Procedure

The method we use to choose $B^R_{\ell'}$ at each step divides DMRG algorithms into two classes, the infinite system procedure and the finite system procedure. In the infinite system procedure, $B^R_{\ell'}$ is chosen to be the spatial reflection of B_ℓ so that $\ell = \ell'$. This means that the size L of the superblock grows by two sites at each iteration. The procedure can be iterated until the energy, calculated in the superblock diagonalization, converges.

The advantage of the infinite system procedure is that calculated quantities scale to their infinite system values. In this sense, this procedure is in the spirit of the original real–space renormalization group. The disadvantages of the infinite system procedure are that for a given system size, it is less accurate than the finite system procedure, and that it cannot easily be generalized to two–dimensional systems. For a two-dimensional system, if a single site is added to the system block at each step, an environment block of the proper geometry cannot in general be formed from the reflected system block.

3.2 The Finite System Procedure

In the finite system procedure, the superblock is formed so that it describes the same finite lattice at each iteration. In other words, the block $B^R_{\ell'}$ is chosen so that $L = \ell + \ell' + 2$ remains fixed. We can do this if we repeat the procedure (which we call a sweep through the lattice) in which the system block is built up from $\ell = 1$ to $\ell = L - 3$ more than once. After one sweep, the system block can be built up from the other side of the lattice, and the stored set of system blocks from that sweep can be used as environment blocks for the next sweep. The procedure is analogous to zipping a zipper back and forth once through the lattice, where the location of the zipper is the location of the single site added to B_ℓ. The sweeps can be repeated until the energy or some other quantity of interest converges. In practice, we have found that it only takes a few sweeps through the lattice to achieve convergence to within truncation error for a given m. The power of this procedure lies in the iterative improvement of the environment block.

On the initial sweep of the finite system procedure, the environment blocks are undefined. For one dimensional systems, however, one can build up the superblock size using the infinite system procedure and use reflections of the stored blocks B_ℓ for $B^R_{\ell'}$ on the initial sweep.

There are a number of advantages to the finite system procedure. First, since the environment blocks are iteratively improved with each sweep through the lattice, the finite system procedure gives much more accurate results for a particular lattice size than the infinite system procedure, although the infinite system procedure can give results that are closer to the thermodynamic limit.

It might be possible to combine the two procedures in a hybrid algorithm to get more accurate results for a given m in the thermodynamic limit.

Second, since the environment block no longer must be a reflection of the system block, it is possible to study lattices that are no longer reflection symmetric. This is useful, for example, in studying systems with impurities or disorder.

Third, in the finite system procedure, the target state of the superblock is the same at each iteration, with unchanging quantum numbers, unlike in the infinite system procedure. For the one-dimensional Heisenberg model calculations described in Ref. [6], the states are labeled only by the z component of the total spin, S_z, so it is easy to find a state with the appropriate quantum number for different lattice sizes. For fermion systems such as the Hubbard model, however, $N_\uparrow$ and $N_\downarrow$, the number of spin up and spin down fermions, are good quantum numbers. Since $N_\downarrow$ and $N_\uparrow$ must be integers, it is impossible to choose them so that the overall occupation stays constant on all different lattice sizes, except at half-filling. The best one can do is to target one or more states closest to the proper density, and this leads to reduced accuracy for non-half-filled systems.

Fourth, it is much easier to extend the finite system procedure to lattices of more than one dimension.

3.3 Extension to Higher Dimensions

One way to extend these algorithms to more than one dimension would be to replace the single sites added between the blocks with a row of sites. However, the extra degrees of freedom added to the system at each real-space blocking would make size of the superblock Hilbert space prohibitively large. Therefore, the two-dimensional algorithms we have developed still involve adding single sites at a time to the system block. This can be done by adding sites in a connected one-dimensional path through the two-dimensional lattice, i.e. by folding the one-dimensional zipper into two dimensions. A typical superblock configuration for the two-dimensional algorithm is shown in Fig. 2. The site added to the system block is enclosed by a dashed line and the dotted line shows the order in which sites are added to the system block for a sweep. One can see that it is not possible to reflect the system block into an environment block of the proper geometry at every iteration, so the finite system algorithm must used. The two-dimensional procedure differs from the one-dimensional finite size procedure only in that there are additional connections between the system and environment blocks along the boundary.

For one-dimensional lattices, we use the infinite system procedure to build up the superblock to the proper size on the first sweep through the lattice. Since this can no longer be done for higher dimensional lattices, we must formulate a procedure for the initial sweep through the lattice. The simplest procedure is to use an empty environment block on the first sweep. One can diagonalize the Hamiltonian for the system block and keep the m states of lowest energy. This procedure is equivalent to Wilson's original numerical renormalization

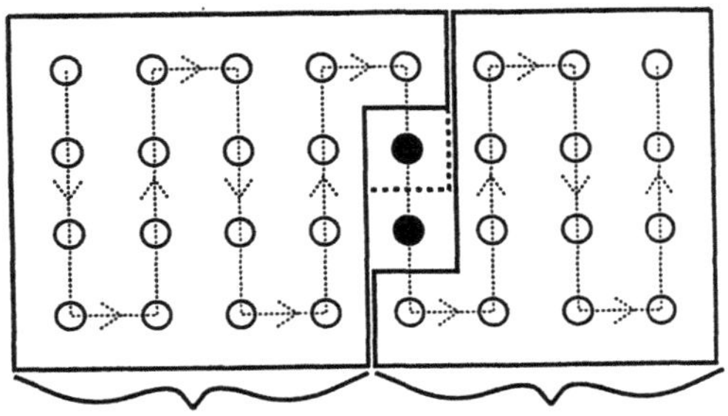

Figure 2: The superblock configuration for the two–dimensional algorithm. The order in which sites are added to the system block on a series of iterations is given by the dotted line, and the site added to the approximate system block Hamiltonian is outline by the dashed line.

group procedure, and is not very accurate even for the one–dimensional single electron on a lattice, as shown in Ref. [5]. In addition, for fermion systems, one must adjust the chemical potential μ so that states with the proper $N_\uparrow$ and $N_\downarrow$ quantum numbers have the lowest energy. The procedure is quite sensitive to these adjustments. Thus, this initialization technique thus tends to be inaccurate and hard to use for fermion systems.

Liang [9] has tried two other techniques for the initial sweep. In the first, he performs an initial infinite system sweep for a one-dimensional lattice, then turns on the additional couplings needed to make the lattice two dimensional on subsequent finite system sweeps. In the second, he uses as the environment block an approximate Hamiltonian for a one–dimensional system of the size of the row length. Both of these procedures depend on representing portions of two-dimensional states by one-dimensional states, and thus give poor representations of the superblock initially.

The technique which we find works best for the initial sweep is a hybrid procedure in which the finite system procedure for a smaller lattice size is repeated for a few iterations, until the system block is big enough so that its reflection can be used for the environment block of a superblock that is a row larger. Thus, the superblock is extended a row at a time. Initially, the first row can be built up with a one-dimensional infinite system procedure. This procedure minimizes problems with target states with inappropriate quantum numbers and provides a reasonable representation for two–dimensional states.

We have found that the accuracy of the initial sweep is not critical as long as the first set of environment blocks has a set of states with appropriate quantum numbers. In most cases, a few sweeps of the finite system procedure will improve the environment blocks sufficiently so that the procedure will converge.

3.4 Performance Considerations

The number of states needed to maintain a certain truncation error in the density matrix projection procedure depends strongly on the number of operators connecting the two parts of the system. Best accuracy is obtained when the number of connections between the system and environment blocks is minimized. Therefore, we study systems with open rather than periodic or antiperiodic boundary conditions. Also, we find that the number of states m needed to maintain a given accuracy depends strongly on the width and weakly on the length of the system.

Just how rapidly the truncation error increases with the width of the system is not clear in general. Liang [9] studied the error in the energy as a function of width for a gas of noninteracting spinless fermions and found that the number of states needed to maintain a given accuracy grew exponentially with the width of the system. In an interacting system such as the Hubbard model, the detailed structure of the energy spectrum seems to be important. For example, in the two chain Hubbard model at half–filling, where there is a spin and pairing gap, the truncation error for a given m is much smaller than away from half–filling, where the spin gap is reduced and the gap to pairing excitations is no longer present. For multiple Hubbard or Heisenberg chains, the presence or absence of a gap in the spin spectrum depends on whether the number of chains is even or odd[10], so the truncation error for a given m depends on the number of chains in a complicated way. Also, increasing the strength of on-site interactions can reduce the truncation error. The Hubbard model DMRG is most accurate for large U and least accurate for $U = 0$.

For systems of more than one dimension, it is therefore important to be able to keep as many states m per block as possible. We have been able to improve the performance of the algorithm in a number of ways. One way of doing this is to minimize the size of the superblock Hilbert space, whose dimension is n^2mm'. For fermion systems, one can reduce the number of states per site n from four to two by treating the spin degree of freedom on the same footing as a spatial coordinate. A site for a particular spatial coordinate and spin can have an occupancy of zero or one fermion. While this makes the path through the lattice (which now has an added dimension) somewhat more complicated, we have found that by adding these "half–sites" instead of full spatial sites on the last few sweeps through the lattice we can increase the accuracy by increasing m. We have also found that m' can be made smaller than m without losing much accuracy in the truncation[6]. Since the representation the approximate block Hamiltonians is poor on the first few sweeps through the lattice, making m large initially does not improve the representation very much. Therefore, the most efficient procedure is to increase m after every sweep through the lattice, so that m' is m from the previous sweep.

We have made a major effort to write the code in an efficient way in C++. We store only the nonzero parts of operators that link states with particular quantum numbers. These matrices are dense in general because the basis transformation at each step mixes matrix elements. This representation minimizes memory usage and makes it possible to optimize highly the multiplication of a vector by the Hamiltonian, the basic step needed for the Lanczos diagonalization. However, the resulting data structures are complicated and are variable in size, so that it has been useful to take advantage of the object-oriented data structures and dynamic memory allocation available in C++. The code is currently limited more by memory usage than by computer time, although we minimize memory usage by writing to disk all operators not needed for a particular superblock diagonalization step. The current version of the code can handle $m = 400$ or more whereas the original Fortran code used in Ref. [6] for the computationally less demanding Heisenberg spin problem could keep at most $m = 200$. We have found that $m \approx 400$ is necessary in order to obtain accurate results for the two-chain Hubbard model away from half-filling.

4 Results for the two-chain Hubbard model

The two-chain Hubbard model is described by the Hamiltonian

$$\begin{aligned} H = \; & -t_y \sum_{i,\lambda\sigma} (c^\dagger_{i,\lambda\sigma} c_{i+1,\lambda\sigma} + c^\dagger_{i+1,\lambda\sigma} c_{i,\lambda\sigma}) \\ & -t_x \sum_{i,\sigma} (c^\dagger_{i,1\sigma} c_{i,2\sigma} + c^\dagger_{i,2\sigma} c_{i,1\sigma}) + U \sum_{i,\lambda} n_{i,\lambda\uparrow} n_{i,\lambda\downarrow}. \end{aligned} \tag{2}$$

We think of the lattice as beinging a ladder aligned with the y axis so that $c^\dagger_{i,\lambda\sigma}$ creates an electron of spin σ at rung j and side $\lambda = 1$ (left) or 2 (right), the hopping along a chain is t_y, the hopping between chains on a rung is t_x, and U is the on-site Coulomb repulsion. This system is thought to be relevant to a number of anisotropic two-dimensional systems, including $(VO)_2P_2O_7$ [11] and $Sr_2Cu_4O_6$ [12, 13], which have weakly coupled ladder-like structures arranged in planes. Here we will concentrate on a parameter regime relevant to the latter class of substances: $U/t_y = 8$, and $t_x = t_y$. We will explore the phase diagram as a function of band filling as the half-filled system is doped with holes.

At half-filling, the Hubbard model maps to the Heisenberg model in the large U/t_y limit. Therefore, the dominant correlations should be antiferromagnetic spin correlations. However, it is known that in the Heisenberg model on two chains [14, 15, 16], there is spin gap leading to an exponential decay of the spin correlation function. The origin of the spin gap is easy to understand in the limit of strong coupling across the rungs. In this case, the only only interaction will be an antiferromagnetic coupling between the two spins on a rung. This two spin system forms a spin singlet state and a higher energy triplet state with an energy separation of the Heisenberg coupling J. Away from half-filling, it is not clear what correlations dominate the behavior. Some authors [17, 18] have predicted that singlet superconductivity with a partial d-wave symmetry should be the dominant order.

In order to resolve these issues, we have calculated equal time spin-spin and pair field correlation functions $S_{\lambda\lambda'}(i,j) = \langle M^z_{i,\lambda} M^z_{j,\lambda'} \rangle$, $D_{xx}(i,j) = \langle \Delta_{xi} \Delta^\dagger_{xj} \rangle$, and $D_{yx}(i,j) = \langle \Delta_{yi} \Delta^\dagger_{xj} \rangle$ with

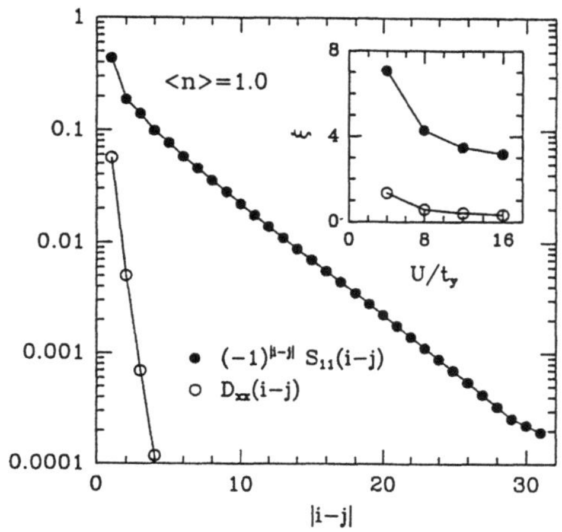

Figure 3: Semilog plot of the spin–spin correlation function $S_{11}(i-j)$ and the pair field correlation function $D_{xx}(i-j)$ at half–filling and $U/t_y = 8$. The insert shows the correlation lengths in units of the lattice spacing obtained from similar plots for various U/t_y values.

$$\begin{aligned} M^z_{i,\lambda} &= n_{i,\lambda\uparrow} - n_{i,\lambda\downarrow} \\ \Delta^\dagger_{xi} &= c^\dagger_{i,1\uparrow}c^\dagger_{i,2\downarrow} - c^\dagger_{i,1\downarrow}c^\dagger_{i,2\uparrow} \\ \Delta^\dagger_{yi} &= c^\dagger_{i+1,2\uparrow}c^\dagger_{i,2\downarrow} - c^\dagger_{i+1,2\downarrow}c^\dagger_{i,2\uparrow}. \end{aligned} \quad (3)$$

Here $S_{11}(i,j)$ and $S_{12}(i,j)$ measure the spin-spin correlations along a chain and between the chains respectively, and $D_{xx}(i,j)$ measures the singlet pair field correlations in which a singlet pair is added at rung j and removed at rung i. In addition, $D_{yx}(i,j)$ measures the pair field correlations in which a singlet pair is added to rung j and removed from the right–hand chain between rungs i and $i+1$. The relative phase of the pair wave function across the ith rung to along one chain from i to $i+1$ is given by comparing the phase of $D_{xx}(i,j)$ to $D_{yx}(i,j)$. This turns out to be negative, corresponding to the mean field result obtained in Ref. [17]. However, the non-interacting $U = 0$ result at a filling $\langle n \rangle = 0.875$ is also negative.

Fig. 3 shows the logarithm of the antiferromagnetic spin–spin correlation function $(-1)^{|i-j|}S_{11}(i-j)$ and the cross–chain pairing correlation function $D_{xx}(i-j)$. Both the correlation functions decay exponentially with $|i-j|$, but the pair field correlations decay much more rapidly. The correlation length, calculated from the slope of the lines in the semilog plot, is plotted as a function of U/t_y in the inset. The spin–spin correlation length decreases as U is increased, saturating at a value near 3 lattice spacing for large U. We have calculated the spin–spin correlation length for the isotropic two chain Heisenberg model using the DMRG [10] and find a value of 3.19 lattice spacings, consistent with the large U limiting value. The pair field correlations decay with a correlation length of the order of a lattice spacing and are thus negligible at half–filling.

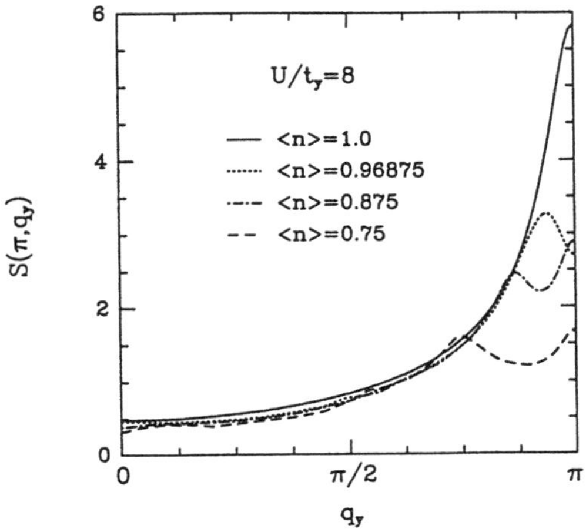

Figure 4: The fourier transform $S(\pi, q_y)$ of the spin–spin correlation function $S_{\lambda\lambda'}(\ell)$. Here $t_x = t_y$ and $U/t_y = 8$ and the calculations were made on a 2×32 lattice.

In order to determine the behavior of the spin correlations as the system is doped below half-filling, we have calculated the magnetic structure factor $S(q_x, q_y)$ by taking the fourier transform of $S_{\lambda\lambda'}(i,j)$. Since the lattice is long in the y direction and the spin–spin correlation function decays exponentially with $|i-j|$, one can take a continuous fourier transform in the y direction without introducing much error. Since there are two chains, q_x can take on the values 0 and π. Only the $S(\pi, q_y)$ branch is interesting, because the correlations are always antiferromagnetic across the rungs. This function is plotted in Fig. 4 for the fillings, $\langle n \rangle = 1.0, 0.9875, 0.875, 0.75$, corresponding to doping 0, 2, 8, and 16 holes into the half–filled 2×32 lattice. As the system is doped away from half-filling, $S(\pi, q_y)$ peaks at a wavevector $q_y = \langle n \rangle \pi$. The residual peak at $q_y = \pi$ present for $\langle n \rangle = 0.875$ and $\langle n \rangle = 0.75$ is present only for even numbers of hole pairs and thus probably disappears in the thermodynamic limit. Therefore, we see that the spin–spin correlations develop incommensurate structure as the system is doped away from half–filling.

One can calculate the spin gap directly, by calculating the difference in energies between the ground state, which has total spin $S = 0$, and the lowest lying $S = 1$ state. We calculate the ground state energy for $N_\uparrow$ spin up electrons and $N_\downarrow$ spin down electrons, $E_0(N_\uparrow, N_\downarrow)$. The spin gap for a system with $N_\uparrow = N_\downarrow = N$ electrons is then given by $\Delta_{spin} = E_0(N+1, N-1) - E_0(N, N)$. The spin gap plotted as a function of filling is shown in Fig. 5. It is largest at half–filling and becomes smaller as the system is doped with holes and seems to be present at least down to fillings of $\langle n \rangle = 0.75$. We show the spin gap for 2×16 and 2×32 lattices to show the size of the finite size effects and argue that they are small enough that the gap is present in the thermodynamic limit for two chains.

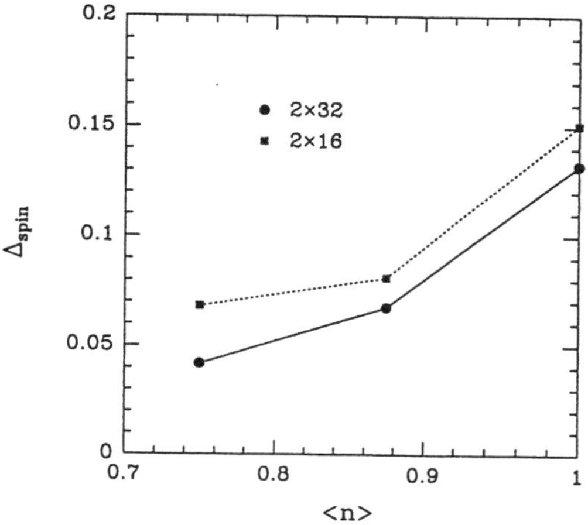

Figure 5: The spin gap Δ_{spin} plotted as a function of band filling $\langle n \rangle$ on a 2×32 lattice for $U/t_y = 8$ and $t_x = t_y$.

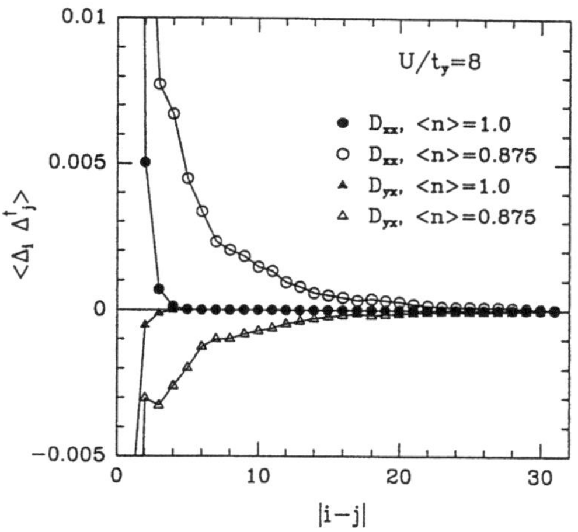

Figure 6: The pair field correlation functions $D_{xx}(i-j)$ and $D_{yx}(i-j)$ versus $|i-j|$ on a 2×32 lattice with $U/t_y = 8$ and $t_x = t_y$.

We now turn to the behavior of the pair field correlations as the system is doped away from half–filling. We have seen that the pairing correlations with cross–chain symmetry decay exponentially in the half–filled system. This is true for all symmetries of the pair field wavefunction. Fig. 6 shows the pair field correlations $D_{xx}(i-j)$ and $D_{yx}(i-j)$ plotted as a function of $|i-j|$ for $\langle n \rangle = 1.0$ and $\langle n \rangle = 0.875$. One can see that $D_{xx}(i-j)$ and $D_{yx}(i-j)$ have opposite signs, as one would expect for d-wave like symmetry, at both fillings and are significantly enhanced for the doped system.

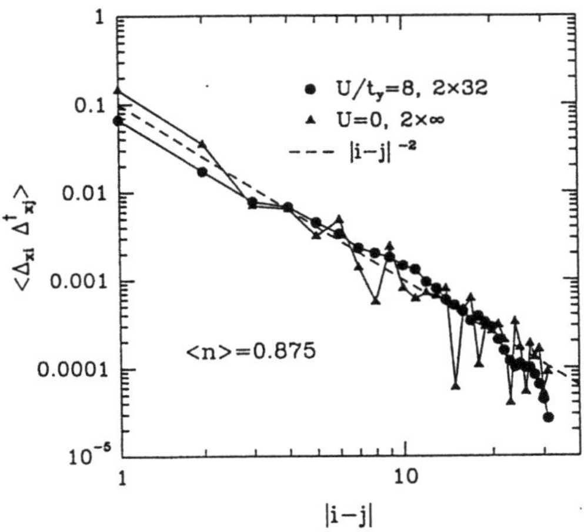

Figure 7: Log-log plot of the rung–rung single pair field correlation function $D_{xx}(i-j)$ versus $|i-j|$ for a 2×32 cluster with $U/t_y = 8$ and an infinite two–chain system with $U = 0$. In both cases $\langle n \rangle = 0.875$. The dashed line shows $|i-j|^{-2}$.

In order to determine the strength of the pairing correlations, one must consider their ℓ–dependence at large distances. For a quasi–one–dimensional system, we expect that any pairing correlation will at best decay as a power of ℓ and can in some cases decay exponentially, as we have seen for the half-filled system. For two chains, one can compare with the the non-interacting $U = 0$ ladder, for which

$$D_{xx}(\ell) = (1/2\pi\ell)^2 \left[2 - \cos(2k_f(0)\ell) - \cos(2k_f(\pi)\ell)\right] . \tag{4}$$

Here $k_f(0) = \cos^{-1}(t_x + \mu)/2$ and $k_f(\pi) = \cos^{-1}(t_x - \mu)/2$ are the Fermi wave vectors corresponding to the bonding and antibonding bands of the two coupled chains with μ the chemical potential. The pair correlations, $D_{xx}(\ell)$ are shown in Fig. 7, plotted on a log–log scale. The correlations of the interacting system decay approximately as ℓ^{-2} and do not seem to be significantly enhanced over those of the non–interacting system, as given by Eq. (4).

5 Conclusion

We have discussed techniques we have developed to apply the density matrix renormalization group to Fermion systems in more than one dimension. In particular, we have been able to obtain accurate results for energy gaps and equal–time correlation functions for the Hubbard model on two coupled chains.

The two–chain Hubbard model is a gapped spin liquid at half–filling. Both spin–spin and pair field correlations decay exponentially, with the spin–spin correlations having the longest correlation length. As the system is doped with holes, the spin–spin correlations become incommensurate at a wave vector proportional to the filling and the spin gap becomes smaller, but persists in the thermodynamic limit. The pairing correlations are enhanced with a *d*-wave–like symmetry and decay algebraically with an exponent close to that of the non–interacting, $U = 0$ system.

Acknowledgements

The authors thank N. Bulut, T.M. Rice, A. Sandvik, M. Vekic, E. Grannan, and R.T. Scalettar for useful discussions. R.M.N. and S.R.W. acknowledge support from the Office of Naval Research under grant No. N00014-91-J-1143 and D.J.S. acknowledges support from the National Science Foundation under grant DMR92–25027. The numerical calculations reported here were performed at the San Diego Supercomputer Center.

References

[1] K.G. Wilson, *Rev. Mod. Phys.* 47, 773 (1975).

[2] J.W. Bray and S.T. Chui, *Phys. Rev.* **B**19, 4876 (1979); S.T. Chui and J.W. Bray, *Phys. Rev.* **B**18, 2426 (1978); J.E. Hirsch,*Phys. Rev.* **B**22, 5259 (1980); C. Dasgupta and P. Pfeuty, *J. Phys.* **C**14, 717 (1981).

[3] P.A. Lee, *Phys. Rev. Lett.* **42** , 1492 (1979).

[4] K.G. Wilson, in an informal seminar.

[5] S.R. White and R.M. Noack, *Phys. Rev. Lett.* **68** 3487, (1992).

[6] S.R. White, *Phys. Rev. Lett.* **69**, 2863 (1992), *Phys. Rev. B* **48**, 10345 (1993).

[7] R.M. Noack, S.R. White, and D.J. Scalapino (to be published).

[8] The term "environment" block is due to Shoudan Liang.

[9] S. Liang (to be published).

[10] S.R. White, R.M. Noack, and D.J. Scalapino (to be published).

[11] D.C. Johnston *et al.*, *Phys. Rev. B* **35**, 219 (1987).

[12] M. Takano, Z. Hiroi, M. Azuma, and Y. Takeda, Jap. J. of Appl. Phys. Series **7**, 3 (1992).

[13] T.M. Rice, S. Gopalan, and M. Sigrist, Europhys. Lett. **23**, 445 (1993).

[14] E. Dagotto, J. Riera, and D.J. Scalapino, *Phys. Rev. B* **45**, 5744 (1992).

[15] T. Barnes et al., *Phys. Rev. B* **47**, 3196 (1993).

[16] S.P. Strong, and A.J. Millis, *Phys. Rev. Lett.* **69**, 2419 (1992).

[17] M. Sigrist, T.M. Rice, and F.C. Zhang (to be published); Sudha Gopalan, T.M. Rice, and M. Sigrist (to be published).

[18] H. Tsunetsugu, M. Troyer, and T.M. Rice (to be published).

Simulation of the Electron-Phonon Interaction in Infinite Dimensions

J.K. Freericks[1] *and M. Jarrell*[2]

[1]Department of Physics, University of California, Davis, CA 95616, USA
[2]Department of Physics, University of Cincinnati, Cincinnati, OH 45221, USA

Abstract. The electron-phonon interaction corresponding to the Holstein model (with Coulomb repulsion) is simulated in infinite dimensions using a novel quantum Monte Carlo algorithm. The thermodynamic phase diagram includes commensurate charge-density-wave phases, incommensurate charge-density-wave phases, and superconductivity. The crossover from a weak-coupling picture (where pairs both form and condense at T_c) to a strong-coupling picture (where preformed pairs condense at T_c) is illustrated with the onset of a double-well structure in the effective phonon potential.

1. Infinite-dimensional formalism

Strong electron-electron correlations are responsible for many important and exotic phenomena in condensed-matter systems including superconductivity, magnetism, heavy fermions, *etc.* Strongly correlated electronic systems are those in which the average electronic correlation energy is equal to or larger than the electronic kinetic energy. Exotic phenomena arise from the competition of simultaneously minimizing the kinetic and potential energy of the electrons. Models of these systems usually do not have analytic solutions. However, recently, Metzner and Vollhardt [1] discovered that these many-body problems simplify in the limit of infinite spatial dimensions. The limit must be taken in such a fashion that the electronic kinetic energy remains finite, so that the effects of the strong electron correlations remain.

Consider the electronic kinetic energy determined by a tight-binding model with hopping between nearest-neighbor sites (with hopping integral t) on a hypercubic lattice in d dimensions. The band structure $\epsilon(\mathbf{k})$ becomes

$$\epsilon(\mathbf{k}) = -2t\sum_{i=1}^{d} \cos \mathbf{k}_i \quad . \tag{1}$$

In the infinite-dimensional limit ($d \to \infty$) the set of $\{\cos \mathbf{k}_i\}$ can be thought of as "random" numbers distributed between -1 and 1 for a general point in the d-dimensional Brillioun zone. The sum of d "random" numbers grows as $\sqrt{d}$, so the band structure remains finite if the hopping integral scales as $t = t^*/2\sqrt{d}$ [1]. Furthermore the central limit theorem states that the density of states corresponding to this band structure [$\rho(y)$] becomes a Gaussian

Springer Proceedings in Physics, Vol. 78
Computer Simulation Studies in Condensed-Matter Physics VII
Editors: D.P. Landau · K.K. Mon · H.-B. Schüttler

distribution

$$\rho(y) = \frac{1}{\sqrt{\pi} t^*} \exp\left(-\frac{y^2}{t^{*2}}\right) \quad , \tag{2}$$

in the infinite-dimensional limit. The number of nearest neighbors ($2d$) diverges, but the hopping between nearest neighbors ($t = t^*/2\sqrt{d}$) vanishes in such a fashion to maintain a finite kinetic energy for the electrons.

The phonon density of states has a very different behavior in the infinite-dimensional limit. The phonon density of states for the Debye model (phonons with a linear dispersion from zero frequency to ω_D) is

$$N(\omega) = \frac{C_d}{\omega_D}\left[\frac{\omega}{\omega_D}\right]^{d-1} \quad , \quad 0 \le \omega \le \omega_D \quad , \tag{3}$$

in d dimensions. In the limit as $d \to \infty$, the phonon density of states becomes a delta function at the Debye frequency.

These two observations for the electron and phonon densities of states motivate one to examine the Holstein-Hubbard model [2, 3] (in which the electrons couple to localized phonons) as the simplest electron-phonon model in infinite-dimensions:

$$\begin{aligned} H &= -\frac{t^*}{2\sqrt{d}} \sum_{(j,k)\sigma} (c^\dagger_{j\sigma} c_{k\sigma} + c^\dagger_{k\sigma} c_{j\sigma}) + \sum_j (g x_j - \mu)(n_{j\uparrow} + n_{j\downarrow} - 1) \\ &+ U_c \sum_j (n_{j\uparrow} - \frac{1}{2})(n_{j\downarrow} - \frac{1}{2}) + \frac{1}{2} M\Omega^2 \sum_j x_j^2 + \frac{1}{2}\sum_j \frac{p_j^2}{M} \quad . \end{aligned} \tag{4}$$

Here $c^\dagger_{j\sigma}$ ($c_{j\sigma}$) creates (destroys) an electron at site j with spin σ, $n_{j\sigma} = c^\dagger_{j\sigma} c_{j\sigma}$ is the electron number operator, and x_j (p_j) is the phonon coordinate (momentum) at site j. The hopping matrix elements connect the nearest neighbors of a hypercubic lattice in d-dimensions and the unit of energy is chosen to be this rescaled matrix element t^*. The phonon has a mass M (chosen to be $M = 1$), a frequency Ω, and a spring constant $\kappa \equiv M\Omega^2$ associated with it. The electron-phonon coupling constant (deformation potential) is denoted by g so that the effective electron-electron attraction becomes the bipolaron binding energy

$$U \equiv -\frac{g^2}{M\Omega^2} = -\frac{g^2}{\kappa} \quad . \tag{5}$$

The Coulomb repulsion is represented by a local Hubbard interaction U_c and the chemical potential is denoted by μ with particle-hole symmetry occurring for $\mu = 0$.

The observation of Metzner and Vollhardt [1] is that the many-body problem also simplifies in the infinite-dimensional limit—both the self energy and the irreducible vertex functions become independent of momentum, and are functionals of the interacting Green's function [1, 4, 5]. The Green's function, self energy, and irreducible vertices still retain their complicated time (frequency) dependence.

The many-body problem is solved by mapping it onto an auxiliary impurity problem [6, 7] in a time-dependent field that mimics the hopping of an electron onto a site at time τ and off the site at a time τ'. The action for the impurity problem is found by integrating out all of the degrees of freedom of the other lattice sites in a path-integral formalism [8]. The lattice is viewed as a reservoir of electrons that can hop onto and off of the local site. Once an electron hops off of the local site, it never returns, because the number of paths that loop through the local site are a factor of $1/d$ smaller than the number of paths that do not loop through the local site. Therefore, the effective action for the impurity problem becomes

$$\begin{aligned} S &= \sum_\sigma \int_0^\beta d\tau \int_0^\beta d\tau' c_\sigma^\dagger(\tau) G_0^{-1}(\tau-\tau') c_\sigma(\tau') \\ &+ \sum_\sigma \int_0^\beta d\tau [gx(\tau)-\mu][n_\sigma(\tau)-\frac{1}{2}] \\ &+ U_c \int_0^\beta d\tau [n_\uparrow(\tau)-\frac{1}{2}][n_\downarrow(\tau)-\frac{1}{2}] + \frac{M}{2}\int_0^\beta d\tau[\Omega^2 x^2(\tau)+\dot{x}^2(\tau)] \end{aligned} \quad (6)$$

where G_0^{-1} is the "bare" Green's function that contains *all of the dynamical information of the other sites of the lattice.* The interacting Green's function, defined to be

$$G(i\omega_n) \equiv \int_0^\beta d\tau e^{i\omega_n \tau} G(\tau) \quad , \quad G(\tau) = -\frac{\mathrm{Tr}\langle e^{-\beta H} T_\tau c(\tau) c^\dagger(0)\rangle}{\mathrm{Tr}\langle e^{-\beta H}\rangle} \, , \quad (7)$$

then satisfies Dyson's equation

$$G_n^{-1} \equiv G^{-1}(i\omega_n) = G_0^{-1}(i\omega_n) - \Sigma(i\omega_n). \quad (8)$$

A self-consistency relation is required in order to determine the bare Green's function G_0. This is achieved by mapping the impurity problem onto the infinite-dimensional lattice thereby equating the full Green's function for the impurity problem with the local Green's function for the lattice

$$\begin{aligned} G_{jj}(i\omega_n) &= \sum_{\mathbf{k}} G(\mathbf{k}, i\omega_n) = \sum_{\mathbf{k}} [i\omega_n + \mu - \epsilon(\mathbf{k}) - \Sigma(i\omega_n)]^{-1} \\ &= F_\infty[i\omega_n + \mu - \Sigma(i\omega_n)]. \end{aligned} \quad (9)$$

Here $F_\infty(z)$ is the scaled complimentary error function of a complex argument.

$$\begin{aligned} F_\infty(z) &\equiv \frac{1}{\sqrt{\pi}} \int_{-\infty}^{\infty} dy \frac{\exp(-y^2)}{z-y} \\ &= -i\mathrm{sgn}[\mathrm{Im}(z)]\sqrt{\pi} e^{-z^2} \mathrm{erfc}\{-i\mathrm{sgn}[\mathrm{Im}(z)]z\}. \end{aligned} \quad (10)$$

The dynamics of the (local) impurity problem is identical to the dynamics of the Anderson impurity model [4, 6, 7, 8, 9]. This many-body problem can be solved exactly with the quantum Monte Carlo (QMC) algorithm of Hirsch and

Fye [10] (see the next section). The impurity is self-consistently embedded in the host, since it must satisfy the self-consistency relation in Eq. (9). Note that this mapping of the infinite-dimensional lattice problem onto a single-site impurity problem is in the *thermodynamic limit*. There are no finite-size effects in infinite-dimensions.

Static two-particle properties are also easily calculated since the irreducible vertex function is local [11]. The static susceptibility for CDW order is given by

$$\begin{aligned}\chi^{CDW}(\mathbf{q}) &\equiv \frac{1}{2N}\sum_{jk\sigma\sigma'} e^{i\mathbf{q}\cdot(\mathbf{R}_j-\mathbf{R}_k)} T\int_0^\beta d\tau \int_0^\beta d\tau' \\ &\qquad [\langle n_{j\sigma}(\tau)n_{k\sigma'}(\tau')\rangle - \langle n_{j\sigma}(\tau)\rangle\langle n_{k\sigma'}(\tau')\rangle] \\ &\equiv T\sum_{mn}\tilde{\chi}^{CDW}(\mathbf{q}, i\omega_m, i\omega_n) = T\sum_{mn}\tilde{\chi}^{CDW}_{mn}(\mathbf{q}) \quad , \end{aligned} \tag{11}$$

at each ordering wavevector $\mathbf{q}$. Dyson's equation for the two-particle Green's function becomes [9, 11]

$$\tilde{\chi}^{CDW}_{mn}(\mathbf{q}) = \tilde{\chi}^0_m(\mathbf{q})\delta_{mn} - T\sum_p \tilde{\chi}^0_m(\mathbf{q})\Gamma^{CDW}_{mp}\tilde{\chi}^{CDW}_{pn}(\mathbf{q}) \quad , \tag{12}$$

with Γ^{CDW}_{mn} the (local) irreducible vertex function in the CDW channel.

The bare CDW susceptibility $\tilde{\chi}^0_n(\mathbf{q})$ in Eq. (12) is defined in terms of the dressed single-particle Green's function

$$\begin{aligned}\tilde{\chi}^0_n(\mathbf{q}) &\equiv -\frac{1}{N}\sum_{\mathbf{k}} G_n(\mathbf{k})G_n(\mathbf{k}+\mathbf{q}) = -\frac{1}{\sqrt{\pi}}\frac{1}{\sqrt{1-X^2}} \\ &\quad \times \int_{-\infty}^{\infty} dy \frac{e^{-y^2}}{i\omega_n + \mu - \Sigma_n - y} F_\infty\left[\frac{i\omega_n + \mu - \Sigma_n - Xy}{\sqrt{1-X^2}}\right]\end{aligned} \tag{13}$$

and all of the wavevector dependence is included in the scalar [6, 12] $X(\mathbf{q}) \equiv \sum_{j=1}^d \cos \mathbf{q}_j/d$. The mapping $\mathbf{q} \mapsto X(\mathbf{q})$ is a many-to-one mapping that determines an equivalence class of wavevectors in the Brillouin zone. "General" wavevectors are all mapped to $X = 0$ since $\cos \mathbf{q}_j$ can be thought of as a random number between -1 and 1 for "general" points in the Brillouin zone. Furthermore, all possible values of X ($-1 \leq X \leq 1$) can be labeled by a wavevector that lies on the diagonal of the first Brillouin zone extending from the zone center ($X = 1$) to the zone corner ($X = -1$). The irreducible vertex function Γ^{CDW}_{mn} is determined by inverting the Dyson equation in Eq. (12) for the *local* susceptibility (which is determined by the Monte Carlo techniques of the following section). Once the irreducible vertex function is found, then Eq. (12) is employed to calculate the momentum-dependent susceptibility.

A similar procedure is used to explore the superconductivity of the model. Here, as in the Hubbard model, it is only necessary to look for superconductivity with the same symmetry as the lattice (s-wave) since other superconductivity with other symmetries do not have pairing interactions [13]. For the singlet s-wave SC channel, the corresponding definitions are as follows: The static susceptibility in the superconducting channel is defined to be

$$\begin{aligned}\chi^{SC}(\mathbf{q}) &\equiv \frac{1}{N}\sum_{jk} e^{i\mathbf{q}\cdot(\mathbf{R}_j-\mathbf{R}_k)} T\int_0^\beta d\tau\int_0^\beta d\tau' \langle c_{j\uparrow}(\tau)c_{j\downarrow}(\tau)c^\dagger_{k\downarrow}(\tau')c^\dagger_{k\uparrow}(\tau')\rangle \\ &\equiv T\sum_{mn}\tilde{\chi}^{SC}(\mathbf{q},i\omega_m,i\omega_n) = T\sum_{mn}\tilde{\chi}^{SC}_{mn}(\mathbf{q}) \quad , \end{aligned} \tag{14}$$

for superconducting pairs that carry momentum $\mathbf{q}$; Dyson's equation becomes

$$\tilde{\chi}^{SC}_{mn}(\mathbf{q}) = \tilde{\chi}^{0}_{m}{}'(\mathbf{q})\delta_{mn} - T\sum_p \tilde{\chi}^{0}_{m}{}'(\mathbf{q})\Gamma^{SC}_{mp}\tilde{\chi}^{SC}_{pn}(\mathbf{q}) \quad , \tag{15}$$

with Γ^{SC}_{mn} the corresponding irreducible vertex function for the SC channel; the bare pair-field susceptibility becomes

$$\begin{aligned}\tilde{\chi}^{0}_{n}{}'(\mathbf{q}) &\equiv \frac{1}{N}\sum_{\mathbf{k}} G_n(\mathbf{k})G_{-n-1}(-\mathbf{k}+\mathbf{q}) = \frac{1}{\sqrt{\pi}}\frac{1}{\sqrt{1-X^2}} \\ &\times \int_{-\infty}^{\infty} dy \frac{e^{-y^2}}{i\omega_n+\mu-\Sigma_n-y} F_\infty\left[\frac{-i\omega_n+\mu-\Sigma^*_n-Xy}{\sqrt{1-X^2}}\right]\end{aligned} \tag{16}$$

with the special value $\tilde{\chi}^{0}_{n}{}'(X=1) = -\mathrm{Im}G_n/\mathrm{Im}(i\omega_n-\Sigma_n)$ for the SC pair that carries no net momentum; and finally the irreducible vertex function is also determined by inverting the Dyson equation in Eq. (15) for the *local* susceptibility.

2. Monte Carlo Algorithm

The dynamics of the impurity problem are identical to that of an impurity embedded in a host metal described by the "bare" Green's function $\mathcal{G}_0$ [4, 8, 9]. Thus, given $\mathcal{G}_0$, the impurity problem may be solved by using the quantum Monte Carlo (QMC) algorithm of Hirsch and Fye [10] (an alternative derivation of this algorithm is presented in the appendix). In the QMC the problem is cast into a discrete path formalism in imaginary time, τ_l, where $\tau_l = l\Delta\tau$, $\Delta\tau = \beta/L$, and L is the number of times slices. The values of L used ranged from 40 to 160, with the largest values of L reserved for the largest values of β because the time required by the algorithm increases like L^3. Since the bare Green's function $\mathcal{G}_0^{-1}$ in Eq. (8) is not *a priori* known, the QMC algorithm must be iterated to determine a self-consistent solution for the Green's function of the infinite-dimensional lattice. The procedure [9] is to begin with a bare Green's function $\mathcal{G}_0^{-1}$, use the QMC algorithm to determine the self energy Σ, calculate the lattice Green's function from Eq. (9), and determine a new bare Green's function from Eq. (8). This process is iterated until convergence is reached (typically $7-9$ iterations). At each step, the precision (the total number of field-configurations generated) is increased. In addition, results from high temperature runs are used to initialize lower temperature runs. These last two steps, are commonly used to to anneal out the state with the lowest free energy.

The details of the (Hirsch-Fye [10]) impurity algorithm, as modified for the Holstein-Hubbard model, are reproduced in the appendix. For the remainder of this section, we will discuss the modifications necessary to apply this algorithm to the infinite-dimensional limit. The main difficulty is that the Hirsch-Fye algorithm requires an imaginary-time path integral technique which only produces data for $G(\tau)$ at a *discrete* set of points in Euclidean time $0 < \tau < \beta$; whereas, the self-consistency step requires either the Matsubara frequency Green's function or the corresponding self energy. This involves a numerical approximation of the integral in Eq. (7). Fourier transforming discretely sampled data presents some well known difficulties [14]. The principle difficulty is that Nyquist's theorem tells us that above some frequency $\omega_n = 1/2\Delta\tau$, unpredictable results are produced by conventional quadrature techniques. Typically this problem is overcome by fitting the discrete data $G(\tau)$ with a smooth cubic spline, and then performing the integral on the splined data [14]. Since the integral on the splined data may be sampled on a much finer grid than the original data, this process is referred to as over sampling.

However, a problem still remains at high frequencies, since the resulting $G(i\omega_n)$ goes quickly to zero for frequencies above the Nyquist cutoff $1/2\Delta\tau$. This presents a difficulty since causality requires that

$$\lim_{\omega_n \to \infty} G(i\omega_n) \sim \frac{1}{i\omega_n} . \tag{17}$$

In order to maintain causality [15] of the Matsubara frequency Green's functions, we condition the Fourier transform with a perturbation theory result. That is, we write

$$G(i\omega_n) = G_{pt}(i\omega_n) + \int_0^\beta d\tau e^{i\omega_n\tau} \left(G(\tau) - G_{pt}(\tau)\right) . \tag{18}$$

where G_{pt} is a Green's function obtained from perturbation theory, and the integral here is performed by the oversampling method described above.

There are two obvious advantages to this approach. First, the integral goes to zero for frequencies greater than the Nyquist frequency $1/2\Delta\tau$, so that the resulting Green's function has the same asymptotic behavior as the perturbation theory result, and is thus causal. Second, often, the perturbation theory result is asymptotically exact (*i. e.* results from a high temperature expansion *etc.*), and this then presents a way of appending exact QMC results at low frequency with asymptotically exact perturbation theory results at high frequency. The flow chart for the resulting algorithm is shown in Fig. 1.

Once convergence of the algorithm in Fig. 1 is reached, the physical properties of the system are calculated with the QMC. A variety of two-particle properties may be calculated in the QMC approach since the irreducible vertex function is also local. For most quantities, this is straight-forward; however, the two-particle Green's functions χ_{mn}^{loc} are difficult to measure efficiently. For example, consider the local opposite-spin particle-particle propagator

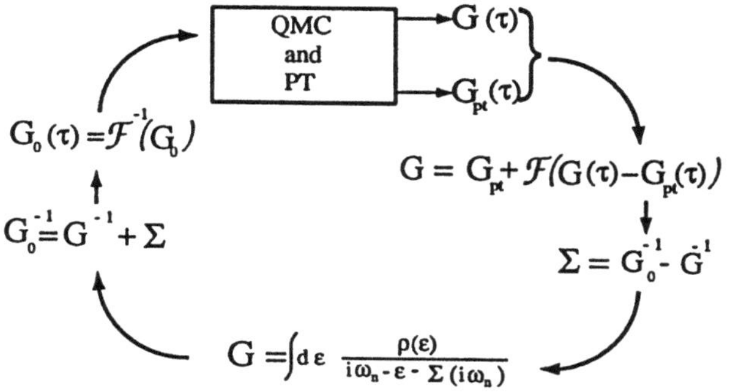

Figure 1. *Flowchart for the $d = \infty$ algorithm. The symbol $\mathcal{F}$ denote that a Fourier transform is to be performed by oversampling, and $\mathcal{F}^{-1}$ denotes its inverse.*

$$\chi_{nm}^{loc} = \int_0^\beta d\tau_1 \cdots d\tau_4 e^{[i\omega_m(\tau_2-\tau_1)-i\omega_n(\tau_3-\tau_4)]} \times \langle T_\tau c_\uparrow(\tau_4) c_\downarrow(\tau_3) c_\downarrow^\dagger(\tau_2) c_\uparrow^\dagger(\tau_1)\rangle . \tag{19}$$

For a particular configuration of the Hubbard-Stratonovich fields, the Fermions are noninteracting, thus the expectation value indicated by the angle brackets above may be evaluated in two steps. First, using Wick's theorem, its value is tabulated for each field configuration $\{s_l, x_l\}$. Second, using Monte Carlo techniques these configurations are averaged over. After the first step, the equation becomes

$$\chi_{nm}^{loc} = \left\langle \int_0^\beta d\tau_1 \cdots d\tau_4 e^{[i\omega_m(\tau_2-\tau_1)-i\omega_n(\tau_3-\tau_4)]} g_\uparrow(\tau_4,\tau_1) g_\downarrow(\tau_3,\tau_2) \right\rangle_{m.c.} \tag{20}$$

where the $m.c.$ subscript means that the Monte Carlo average is still to be performed.

To measure this on the computer, the integrals must be approximated by sums. Since the Green's functions change discontinuously when the two time arguments intersect, the best integral approximation that can be used here is the trapezoidal approximation. Using this, we will run into Green's functions with both time arguments the same $g(j,j)$. This is stored as $g(j^+,j)$ (i.e. it is assumed that the first time argument is slightly greater than the second), but in the sums we clearly want the equal time Green's function to be the average $\{g(j^+,j) + g(j,j^+)\}/2 = g(j^+,j) - 1/2$. If we call g, with 1/2 subtracted from its diagonal elements, $\overline{\mathbf{g}}$, then

$$\chi_{nm}^{loc} = \left\langle \left(\sum_{j,k} \Delta\tau \; e^{+i\pi j(2n+1)/L} \overline{g}_{\uparrow}(j,k) \; \Delta\tau \; e^{-i\pi k(2m+1)/L} \right) \right. \tag{21}$$

$$\left. \left(\sum_{p,q} \Delta\tau \; e^{-i\pi p(2n+1)/L} \overline{g}_{\downarrow}(p,q) \; \Delta\tau \; e^{+i\pi q(2m+1)/L} \right) \right\rangle_{m.c.}$$

This measurement may be performed efficiently if each term in parenthesis is tabulated first and stored as a matrix, and then the direct product of the two matrices taken as the estimate of χ^{loc}. When done this way, the time required for this measurement scales like $\sim L^3$ rather than $\sim L^4$ as would result from a straight-forward evaluation of the sums implicit in Eq. (20).

Finally, the irreducible vertex function is determined by inverting the relevant local Dyson equation. The momentum-dependent susceptibility may then be calculated from Eq. (12) or (15).

3. Results

For the results presented here, we chose an intermediate phonon frequency $\Omega = 0.5t^*$ (which is approximately one-eight of the effective electronic bandwidth) for which there is a competition between CDW and SC order. As shown in Fig. 3, CDW order is favored near half filling (due to Fermi surface nesting) and SC order is favored away from half filling. As shown in Fig. 2b, there is a maximum CDW transition temperature, because it decreases as the coupling strength increases in the strong-coupling regime.

In order to shed some light on the transition from weak to strong coupling the QMC simulations were sampled to determine a time-averaged effective phonon potential. The probability $P(x)$ that the phonon coordinate $x(\tau_l)$ lies in the interval from x to $x + \delta x$ was calculated for each time slice τ_l and averaged over all time slices. An effective phonon potential $V_{eff.}(x)$ was then extracted from the probability distribution $P(x) \propto \exp[-\beta V_{eff.}(x)]$ [17]. This effective potential is plotted in Fig. 2 (a) for four different values of the electron-phonon coupling strength at a temperature $T = 1/7$. In the case of weak coupling ($g = 0.325$), the potential appears harmonic. The potential flattens when $U \approx t^*$ ($g = 0.5$) and as g increases further, a double-well structure develops [18]. The barrier height grows linearly with g as does the separation of the minima. The peak of the $T_c(g)$ curve for the CDW transition [see Fig. 2(b)] is reached at the point where the barrier height is on the order of T_c ($g = 0.625$). Beyond this point ($g = 1.0$) the system enters the strong-coupling regime and T_c decreases.

In the region where the double-well potential has developed, the phonon coordinate tunnels between the wells and the tunneling rate decreases as the temperature is lowered below the barrier height. At this point the system may be considered to be a random mixture of empty sites and bipolarons that fluctuates in time. Tunneling through the barrier produces correlations

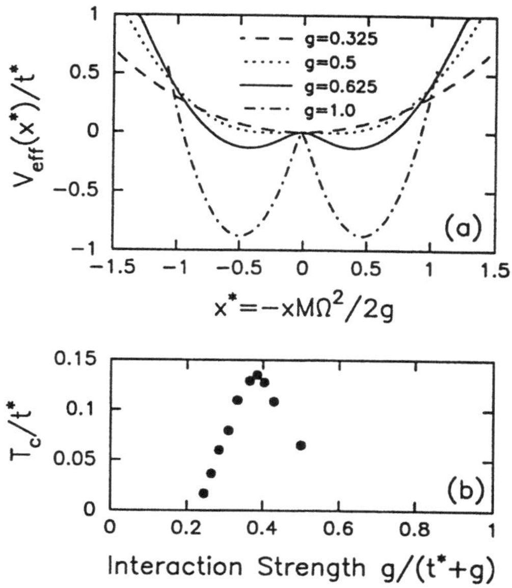

Figure 2: *(a) Effective potential for the phonon (after integrating out the effects of the electrons) as a function of electron-phonon coupling and (b) the CDW transition temperature at half filling as a function of the coupling. The parameters chosen here are $\Omega = 0.5t^*$ and $U_c = 0$. Note that the maximum in the T_c curve occurs when the barrier height of the double-well potential is equal in magnitude to T_c.*

between the empty-sites and the bipolarons resulting in a condensed CDW phase. However as the barrier height increases, the transition temperature drops because the tunneling is suppressed. The transition temperature reaches its maximum at the point where the barrier height is equal in magnitude to T_c.

As the system is doped away from half-filling there is a competition between CDW order and superconductivity . We find that the CDW-ordered state remains "locked" at the "antiferromagnetic" point ($X = -1$) for a wide range of dopings away from half-filling. Figure 3 displays the results for the transition temperature of the Holstein-Hubbard model as a function of electron concentration for two different values of U_c at $g = 0.5t^*$. In the case where $U_c = 0$ the system must be doped out to a concentration of $\rho_e = 0.52$ before it becomes superconducting. There was no evidence for any incommensurate order when $U_c = 0$. We expect that a Coulomb repulsion will favor the SC phase over the CDW phase because the Coulomb repulsion directly reduces the CDW interaction, but is not as effective at reducing the SC interaction because of the retardation between the pairing electrons which allows the electrons to attract each other without being at the same site at the same time (the so-called pseudopotential effect). This result is clearly seen in Figure 3, where the phase space for the SC order increases when $U_c = 0.5$. Note that a

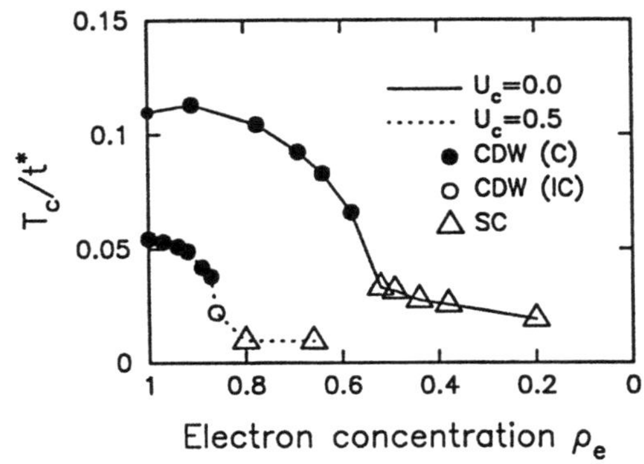

Figure 3: *Transition temperature for the Holstein-Hubbard model at $\Omega = 0.5t^*$ and $g = 0.5t^*$. As the Coulomb repulsion is increased, the SC phase becomes more stable, as do incommensurate CDW phases.*

finite Coulomb repulsion also favors the appearance of incommensurate phases, which now can be detected with the QMC techniques.

Acknowledgments

We would like to acknowledge useful conversations with H. Akhlaghpour, D. L. Cox, R. M. Fye, Th. Pruschke, R. Scalettar, and D. J. Scalapino. This work was supported by the National Science Foundation grant number DMR-9107563 and by the Office of Naval Research under Grant No. N00014-93-1-0495. In addition MJ would like to acknowledge the support of the NSF NYI program. Computer support was provided by the Ohio Supercomputer Center.

Appendix. Derivation of Quantum MC algorithm of Hirsch and Fye

The purpose of this section is to derive the Hirsch-Fye algorithm [10] using Grassmann algebra. We begin by splitting the single impurity Anderson model Hamiltonian, into bare and interacting parts, $H = H_0 + H_1 + H_2$, where

$$\begin{aligned} H_0 &= \sum_{k\sigma} \epsilon(\mathbf{k}) c^\dagger_{k\sigma} c_{k\sigma} + V \sum_{k\sigma} (f^\dagger_\sigma c_{k\sigma} + c^\dagger_{k\sigma} f_\sigma) \\ &+ (gx - \mu)(n_{f\uparrow} + n_{f\downarrow} - 1) + \frac{1}{2} M\Omega^2 x^2 \quad , \end{aligned} \tag{22}$$

$$H_1 = U(n_{f\uparrow} - 1/2)(n_{f\downarrow} - 1/2)\,, \tag{23}$$

and

$$H_2 = \frac{p^2}{2M}\,. \tag{24}$$

To obtain the Trotter-Suzuki decomposition for the partition function [19] we divide the interval $[0, \beta]$ into L sufficiently small subintervals such that the commutators $\Delta\tau^2 [H_0, H_I]$ *etc.* may be neglected. This leads to

$$Z = Tre^{-\beta H} = Tr \prod_{l=1}^{L} e^{-\Delta\tau H} \approx Tr \prod_{l=1}^{L} e^{-\Delta\tau H_0} e^{-\Delta\tau H_1} e^{-\Delta\tau H_2} . \tag{25}$$

The interacting part of the Hamiltonian, H_1, may be further decoupled by mapping it to an auxiliary Ising field via a discrete Hirsch-Hubbard-Stratonovich [20] transformation,

$$e^{-\Delta\tau H_I} = e^{-\Delta\tau U(n_{f\uparrow}-1/2)(n_{f\downarrow}-1/2)} = \frac{1}{2} e^{-\Delta\tau U/4} \sum_{s=\pm 1} e^{\alpha s(n_{f\uparrow}-n_{f\downarrow})} \tag{26}$$

where $\cosh(\alpha) = e^{\Delta\tau U/2}$. Finally, one may cast Eq. (25) into functional-integral form by using coherent states [Grassmann variables for Fermions, and complex numbers for the Bosons, $a = \sqrt{m\Omega/2}(x + ip/m\Omega)$ and a^*]. If we integrate out the host Fermionic degrees of freedom $\{c_{k\sigma}\}$ as well as the momentum of the phonon, then we end up with

$$S_{eff} = (\Delta\tau V)^2 \sum_{l,l',\sigma} f^*_{\sigma,l} G_0(l,l') f_{\sigma,l'} + S_{int} + S_B , \tag{27}$$

where

$$S_B = \frac{\Delta\tau}{2} \sum_l \left[\left(\frac{x_l - x_{l+1}}{\Delta\tau} \right)^2 \right] + \Omega^2 x_l^2 , \tag{28}$$

$$\begin{aligned} S_{int} &= \sum_{l,\sigma,x_l} -f^*_{\sigma,l}(f_{\sigma,l} - f_{\sigma,l-1}) \\ &+ \Delta\tau f^*_{\sigma,l}(g x_l - \mu + \frac{U}{2} + \frac{\alpha}{\Delta\tau} s_l \sigma) f_{\sigma,l-1} \quad , \end{aligned} \tag{29}$$

and

$$G_0^{-1}(l,l') = \frac{1}{N} \sum_k \delta_{l,l'} - \delta_{l-1,l'}[1 - \Delta\tau\epsilon(\mathbf{k})] . \tag{30}$$

At this point the correspondence of the impurity and the infinite-dimensional Hubbard model is clear. In both, G_0 contains the information about the host into which the impurity is embedded. The difference is that G_0 must be determined self-consistently for the lattice model.

We will now proceed to derive the Monte Carlo algorithm [10] sufficient for either the impurity or the infinite-dimensional lattice problem. By integrating over $f_{\sigma,l}$ we can write down the partition function (neglecting a numerical prefactor), as

$$Z = \sum_{\{s_l,x_l\}} det(G^{-1}_{\uparrow\{s_l,x_l\}}) det(G^{-1}_{\downarrow\{s_l,x_l\}}) e^{-S_B} \tag{31}$$

where

$$G^{-1}_{\sigma,\{s_l,x_l\}}(l,l') = \delta_{l,l'} - \delta_{l-1,l'}[1-\Delta\tau\,(gx_l-\mu)+\alpha s_l\sigma] - \Delta\tau^2V^2G_0(l,l') \quad (32)$$

and we sum over all configurations of Hubbard-Stratonovich and phonon fields $\{s_l, x_l\}$. If we reexponentiate the above formula by defining $\mathcal{V}_{\sigma,\{s_l,x_l\}}(l) \equiv \Delta\tau(gx_l - \mu + \alpha s_l\sigma/\Delta\tau)$, we can write it in a simple notation as

$$G_\sigma^{-1} = 1 + Te^{\mathcal{V}_\sigma} + \Delta\tau^2V^2G_0 \; , \quad (33)$$

where T is $\delta_{l-1,l'}$ and $\mathcal{V}_\sigma \equiv \mathcal{V}_{\sigma,\{s_l,x_l\}}(l)$ for one special configuration. For another field configuration the only difference comes from $\mathcal{V}_\sigma$ such that $G'^{-1}_\sigma - G^{-1}_\sigma = T(e^{\mathcal{V}'_\sigma} - e^{\mathcal{V}_\sigma}) + O(\Delta\tau^{3/2})$ (note that α is of the order of $\Delta\tau^{1/2}$). On the other hand $T = (G^{-1}_\sigma - 1 - \Delta\tau^2V^2G_0)e^{-\mathcal{V}_\sigma}$ which results in

$$G'^{-1}_\sigma - G^{-1}_\sigma = (G^{-1}_\sigma - 1)e^{-\mathcal{V}_\sigma}(e^{\mathcal{V}'_\sigma} - e^{\mathcal{V}_\sigma}) + O(\Delta\tau^{3/2}) \, . \quad (34)$$

Multiplying from the left by G and from the right by G' and, ignoring terms $O(\Delta\tau^{3/2})$, we find

$$G'_\sigma = G_\sigma + (G_\sigma - 1)(e^{\mathcal{V}'_\sigma - \mathcal{V}_\sigma} - 1)G'_\sigma \; , \quad (35)$$

or

$$G_\sigma G'^{-1}_\sigma = 1 + (1 - G_\sigma)(e^{\mathcal{V}'_\sigma - \mathcal{V}_\sigma} - 1) \quad . \quad (36)$$

The probability of having a configuration $\{s_l, x_l\}$ is $P_{sx} \propto det(G^{-1}_{\uparrow\{s_l,x_l\}}) \times det(G^{-1}_{\downarrow\{s_l,x_l\}})e^{-S_B}$; on the other hand the detailed balance requires

$$P_{sx'}P_{sx'\rightarrow sx} = P_{sx}P_{sx\rightarrow sx'} \quad , \quad (37)$$

for all sx'. We may satisfy this requirement by defining the probability of going from $\{s_l, x_l\}$ to $\{s'_l, x'_l\}$ as $R/(1+R)$, where

$$R \equiv \frac{det(G'_\uparrow)det(G'_\downarrow)e^{-S'_B}}{det(G_\uparrow)det(G_\downarrow)e^{-S_B}} \quad (38)$$

is the relative weight of two configurations. If the difference between two configuration is due to a flip of a single Hirsch-Hubbard-Stratonovich field at the mth imaginary time slice then [10]

$$R = \prod_\sigma [1 + (1 - G_{\sigma m,m})(e^{-2\alpha\sigma s_m} - 1)] \, , \quad (39)$$

or, if the difference is due to a change in the phonon displacement $x_l \rightarrow x'_l$, then

$$R = e^{S'_B - S_B}\prod_\sigma [1 + (1 - G_{\sigma m,m})(e^{\Delta\tau g(x_l - x'_l)} - 1)] \, . \quad (40)$$

Finally we can write down the evolution of the Green's function in the QMC time, when for example we flip a single Hirsch-Hubbard-Stratonovich field at the mth imaginary time slice [10]

$$\begin{aligned} G'_{\sigma i,j} &= G_{\sigma i,j} + (G_{\sigma i,m} - \delta_{i,m})(e^{-2\alpha\sigma s_m} - 1) \\ &\quad \times\{1 + (1 - G_{\sigma m,m})(e^{-2\alpha\sigma s_m} - 1)\}^{-1} G_{\sigma m,j} \,. \end{aligned} \tag{41}$$

Similarly, when we change a single boson x_m at the mth time slice,

$$\begin{aligned} G'_{\sigma i,j} &= G_{\sigma i,j} + (G_{\sigma i,m} - \delta_{i,m})(e^{\Delta\tau g(x_l - x'_l)} - 1) \\ &\quad \times\{1 + (1 - G_{\sigma m,m})(e^{\Delta\tau g(x_l - x'_l)} - 1)\}^{-1} G_{\sigma m,j} \,. \end{aligned} \tag{42}$$

The QMC process precedes by sequentially proposing changes in each field, accepting these changes with probability $P_{sx->sx'}$, and updating the Green's function with Eq. (41) or Eq. (42) when the change is accepted. In addition to using local moves in which a single spin or a single phonon field is changed, we also employ global moves, in which either all of the spins are flipped, or all of the phonon coordinates are shifted,

References

[1] W. Metzner and D. Vollhardt, Phys. Rev. Lett. **62**, 324 (1989).

[2] T. Holstein, Ann. Phys. **8**, 325 (1959).

[3] J. Hubbard, Proc. Royal Soc. London (Ser. A) **276**, 238 (1963).

[4] H. Schweitzer and G. Czycholl, Z. Phys. B **77**, 327 (1990).

[5] W. Metzner, Phys. Rev. B **43**, 8549 (1991).

[6] U. Brandt and C. Mielsch, Z. Phys. B **75**, 365 (1989).

[7] F. J. Ohkawa, Phys. Rev. B **44**, 6812 (1991); Prog. Theor. Phys. Suppl. **106**, 95 (1991).

[8] A. Georges and G. Kotliar, Phys. Rev. B **45**, 6479 (1992).

[9] M. Jarrell, Phys. Rev. Lett. **69**, 168 (1992).

[10] J. E. Hirsch and R. M. Fye, Phys. Rev. Lett. **56**, 2521 (1986).

[11] V. Zlatić and B. Horvatić, Solid State Commun. **75**, 263 (1990).

[12] E. Müller-Hartmann, Z. Phys. B **74**, 507 (1989); Z. Phys. B **76**, 211 (1989).

[13] M. Jarrell and Th. Pruschke, Z. Phys. B. **90**, 187(1993).

[14] W. H. Press, B. P. Flannery, S. A. Teukolsky, and W. T. Vetterling, *Numerical Recipes* (Cambridge University Press, Cambridge, 1986).

[15] Obeying causality is crucial to guarantee that the procedure described here is properly Herglotz.

[16] J. K. Freericks, M. Jarrell, and D. J. Scalapino, Phys. Rev. B **48**, 6302 (1993); *to appear in* Europhys. Lett. (1994)

[17] We would like to thank H.-B. Schüttler for suggesting we examine an effective phonon potential in this manner. See also J. Mustre de Leon, *et. al.* Phys. Rev. Lett. **68**, 3236 (1992); J. Zhong and H.-B. Schüttler, Phys. Rev. Lett. **69**, 1600 (1992).

[18] These results are qualitatively similar to those found in C. C. Yu and P. W. Anderson, Phys. Rev. B **29**, 6165 (1984).

[19] H. F. Trotter, Proc. Am. Math. Soc. **10** 545 (1959); M. Suzuki, Prog. Theor. Phys. **56**, 1454 (1976)

[20] J. E. Hirsch, Phys. Rev. B, **28**, 4059 (1983); **31**, 4403 (1985).

Excitation Spectra in the 1-D Hubbard Model from Quantum-Monte-Carlo Simulations

A. Muramatsu, R. Preuss, W. von der Linden, P. Dieterich, F.F. Assaad, and W. Hanke

Physikalisches Institut, Universität Würzburg, Am Hubland, D-97074 Würzburg, Germany

Abstract. The numerical simulation of quantum mechanical many-body systems by Monte Carlo Methods is in general only able to deliver ground-state or thermodynamical expectation values of physical observables. Recent developments however, made possible to perform an analytic continuation of imaginary-time quantum Monte Carlo data into real-frequency spectra. In particular the maximum-entropy method (MEM) was successfully applied to the Anderson impurity model and the Heisenberg quantum antiferromagnet both in one (1-D) and two (2-D) dimensions. We present here an application of the MEM on quantum Monte Carlo (QMC) simulations of the 1-D Hubbard model for large system sizes ($N \leq 84$), such that the characterization of dispersion relations for excitation spectra becomes possible. The one-particle excitations posses cosine-like bands that, surprisingly, agree extremely well with slave-boson mean-field ones. Further comparisons with exact results from Bethe-Ansatz and conformal field-theory demonstrate the reliability of both the QMC simulations as well as the MEM.

1 Introduction

A considerable amount of progress was achieved in the last ten years on numerical methods that allow to study strongly correlated fermionic lattice-models in an essentially exact way. Besides exact diagonalizations (ED) [1] that are limited to rather small systems (~ 20 sites) and, thus, in general not able to perform finite-size scaling studies, quantum Monte Carlo (QMC) methods both for finite temperature [2] as well as for $T = 0$ [3, 4] emerged as reliable tools, in principle not subjected to the above mentioned limitations. The QMC simulations however, are designed to work in imaginary time such that only ground-state expectation values or thermodynamic averages of static observables are directly delivered by these methods.

Recent developments towards extracting dynamical information from the QMC data [5, 6, 7, 8, 9, 10] concentrated in the determination of spectral functions from the corresponding imaginary-time correlation functions, where the maximum-entropy method [8, 9, 10] offers at the moment the best way for a controlled analytic continuation to real-frequency data. This method was successfully applied to the single impurity Anderson model [11, 9], the 1-D [12] and 2-D [13] Heisenberg antiferromagnet, and the 2-D Hubbard model [10, 14].

Parallel to the above developments in numerical methods, a strong impulse for understanding strongly correlated systems originated from the field of high temperature superconductivity, where a number of experimental results [15] show that the cuprate superconductors are metals where electronic correlation is dominant. In particular, Anderson [15] proposed that the high-T_c cuprates (HTC) should be considered as Luttinger

Springer Proceedings in Physics, Vol. 78
Computer Simulation Studies in Condensed-Matter Physics VII
Editors: D.P. Landau · K.K. Mon · H.-B. Schüttler

liquids [16], characterized by elementary excitations that cannot be directly related to the bare electrons as the quasi-particles in a Fermi-liquid. The best known Luttinger liquid is the 1-D Hubbard model. In spite of the fact that it can be solved exactly by Bethe-Ansatz (BA) [17], very little was known about their dynamical properties until recently. A number of authors succeeded in extracting from BA information about spectral properties [18, 19, 20, 21, 22], however, many of these results are limited to special situations (e.g. half-filling, $U \to \infty$ and/or one-hole doping, or $\omega = 0$). Only recent progress achieved in the frame of conformal field-theory [23], led to the asymptotic properties of correlation functions irrespective of the coupling constant and doping. In spite of the importance of these developments, a general description of spectral properties at finite frequencies is still lacking.

The present work is mainly concerned with the application of QMC simulations and the extraction of dynamical data for a Luttinger-liquid, namely the 1-D Hubbard model. Simulations were performed for rather large systems (up to 84 sites) in order to characterize the dispersion relations of one-particle, spin- and charge-density excitations. The findings are described in the following [24].

For half-filling, where it is known that the system is a Mott-insulator, two bands appear that agree extremely well with the slave-boson mean-field (SBMF) [25] dispersion obtained in the antiferromagnetic state. The spectral weight, however, is not evenly spread between the two bands as in the mean-field ones. Rather, it is observed that weight is transferred from the lower to the upper band as one approaches the Fermi-wavevector k_F from below (i.e. $k < k_F$) until for $k \gg k_F$ most of the weight is on the upper band. Such a continuous process of transfer of spectral weight is certainly beyond the capability of any mean-field approximation. When the system is doped, a single band is observed, again in very close agreement with SBMF results in the paramagnetic state. Thus, it appears as a general feature, that the SBMF bands are able to accurately describe the dispersion of the one-particle excitations in spite of the Luttinger-liquid character of the 1-D Hubbard model and the SBMF being a Fermi-liquid. This situation is reminiscent of the HTC, where band-structure calculations [26] agree well with angle-resolved photoemission results [27], while other measurements seem to signal a departure from a Fermi-liquid [15]. A further interesting feature related to the one-particle excitations is shown by the density of states, where a depression appears at the Fermi-energy. We believe that it signals the fact that in a Luttinger liquid the density of states vanishes at the Fermi-energy as $\sim |\omega|^\alpha$, where α is a power determined by the coupling constant U of the model and by the density of particles. Since for the system considered here, $\alpha \sim 0.04$ (see Sec. 3), the vanishing itself of the density of states cannot be observed for a realistic resolution but only a depression remains. This is in sharp contrast to results obtained recently by similar methods on the 2-D Hubbard model [14], where a peak develops near that region, as it was found in simulations of the infinite-dimensional Hubbard model. It appears therefore questionable for the 2-D Hubbard model to be a Luttinger-liquid.

The accuracy of the method is tested by comparison with known exact results for spin- and charge-density excitations. At half-filling, both the "spinon"-velocity and the full dispersion relation for spin excitations agree with results extracted from BA [18]. In the doped case, the spinon and holon velocities obtained lead to a value of the conformal charge of $c = 0.98 \pm 0.05$ for $N = 84$ sites, in very good agreement with the exact value $c = 1$ [23]. The spectrum for spin- and charge-excitations fulfills excellently a frequency sum rule [28] for each k-point except for those close to $2k_F$, where departures of at most 10% are observed. The wide range of spectral properties that very accurately agree with exact results demonstrates the reliability of the applied methods.

The model and the numerical methods used are described shortly in the next section. In Section 3 we display and discuss the results obtained so far.

2 Model and numerical methods

We consider the 1-D Hubbard model described by the following Hamiltonian:

$$H = -t\sum_{i,\sigma}\left(c^{\dagger}_{i+1,\sigma}c_{i,\sigma} + h.c.\right) + U\sum_{i} n_{i\uparrow}n_{i\downarrow} , \qquad (1)$$

where $c^{(\dagger)}_{i,\sigma}$ are annihilation (creation) operators for an electron at site i with spin σ, and $n_{i\sigma} = c^{\dagger}_{i,\sigma}c_{i,\sigma}$. The first term describes the kinetics of spin-$\frac{1}{2}$ fermions in the tight-binding approximation, where they can be thought of electrons on s-orbitals that only overlap with the nearest neighbors. The second term is an extreme crude representation of the Coulomb repulsion. It has a value U within one orbital and is set to zero for fermions on different sites. In spite of its simplicity, perturbative methods have failed to describe correctly the underlying physics. An exact solution that allows to calculate the ground-state energy and several thermodynamic quantities was achieved by Lieb and Wu in the frame of Bethe-Ansatz [17]. This solution however, does not allow for the calculation of general correlation functions and in particular of excitation spectra. As mentioned in the introduction, several spectral properties were extracted from Bethe-Ansatz like the dispersion of spin-excitations [18], that will be used below in order to test the accuracy of our results. The 1-D Hubbard model is therefore the ideal ground to test the numerical methods but at the same time to gain further physical insight in the system.

The simulations were performed in the grand canonical ensemble with finite temperature [29]. Although several articles describe in detail the algorithm due to Blankenbecler, Scalapino and Sugar (BSS) [30, 31, 32], we describe below a few features that clarify the notation.

In order to compute the partition function of the system,

$$Z = \mathrm{Tr}e^{-\beta H} = \mathrm{Tr}\left(e^{-\Delta\tau H}\right)^{L} , \qquad (2)$$

where $\beta = (k_B T)^{-1}$ is the inverse temperature and k_B the Boltzmann factor, the exponential is expressed as a product of L terms with $\Delta\tau = \beta/L$. Then for each factor, the following (Trotter) approximation can be made,

$$e^{-\Delta\tau H} \simeq e^{-\Delta\tau H_1}e^{-\Delta\tau H_2} , \qquad (3)$$

where H_1 and H_2 correspond to the first and second terms in eq. (1) respectively, and an error of $\mathcal{O}(\Delta\tau^2)$ was made. At this stage, the interaction term can be decoupled by means of a discrete Hubbard-Stratonovich transformation [33], where auxiliary Ising-like classical field are introduced:

$$e^{-\Delta\tau U n_{\uparrow}n_{\downarrow}} = \frac{1}{2}\sum_{s} e^{\lambda s(n_{\uparrow}-n_{\downarrow}) - \frac{1}{2}\Delta\tau U(n_{\uparrow}+n_{\downarrow})} , \qquad (4)$$

where $\cosh\lambda = \exp(\Delta\tau U/2)$ and $s = \pm 1$. Once the interaction term is replaced by a bilinear form in the fermion operators, the trace over fermionic degrees of freedom can be performed, giving a fermionic determinant det$[s]$ for each configuration $\{s\}$ [31]. The modulus of the determinant will serve as a weight and the sum over the different configurations $\{s\}$ is made using importance sampling. Therefore, the expectation values of an observable will have essentially the following form,

$$< \hat{O} >= \frac{\sum_{\{s\}} \hat{O}\mathrm{sign}|\det[s]|}{\sum_{\{s\}} \mathrm{sign}|\det[s]|} . \tag{5}$$

Here we used sign $\equiv$ sign(det[s]) such that a positive definite weight is obtained for the Monte Carlo sampling whereas the sign of the determinant is followed as a part of the observable. Since the average sign enters in the denominator of eq. (5), large fluctuations are found in the case that the average sign goes to zero, a situation that frequently appears for fermionic simulations and is termed the minus-sign problem [34]. A recent review by D.J. Scalapino [2] gives an overview of the applications of this algorithm for different aspects of the Hubbard model.

We have performed simulations of the 1-D Hubbard model with a number of sites N ranging from 12 to 84 with periodic boundary conditions. In all the cases, the number of particles N_p was chosen to be $N_p = 4k + 2$, where k is an integer, such that the minus sign problem does not affect the simulations. In order to be able to study the low-energy region of the model, the simulations were carried out at inverse temperatures as low as $\beta = 20/t$, where $\Delta\tau$, was between 0.1 and 0.125. In all the cases, the interaction strength was $U = 4t$, that in one dimension equals the bandwidth of non-interacting fermions, and therefore corresponds to a strong interaction.

We close this section by succintly describing the maximum-entropy method used to perform the analytic continuation of the data in imaginary time obtained in the QMC simulations to real frequencies spectra. For a complete and detailed account of the method we refer to the original works [8, 9]. As an example we consider the one-particle Green's function in imaginary time $G_\sigma(\vec{k},\tau) =< c_{\vec{k},\sigma}(\tau)c_{\vec{k},\sigma}\sigma^\dagger >$ and its relation to the spectral function given by

$$G(\vec{k},\tau) = \int_{-\infty}^{\infty} d\omega \frac{e^{-\omega\tau}}{1+e^{-\beta\omega}} A(\vec{k},\omega) , \tag{6}$$

where a sum over σ is implicit in the last equation. In order to obtain $A(\vec{k},\omega)$ eq. (6) should be inverted, that amounts to an inverse Laplace transform. This is an ill-posed problem in the presence of statistical error in the QMC data $G(\vec{k},\tau)$. In the following we describe the steps followed in order to obtain the spectra without going into the general frame of probability theory upon which maximum-entropy relies [8, 9].

First, a discretized form of eq. (6) is proposed,

$$\bar{G}(\vec{k},\tau_j) = \sum_i \Delta\omega A(\vec{k},\omega_i) \frac{e^{-\omega_i\tau_j}}{1+e^{-\beta\omega_i}} , \tag{7}$$

where τ_j correspond to the L points on the imaginary-time axis β at which the QMC data are calculated and ω_i give a corresponding mesh on the real frequency axis. In the case of the one-particle Green's function it is understood that the frequencies are measured from the chemical potential. Assuming that the data $G(\vec{k},\tau_i)$ are Gaussian distributed with error σ_i, a probability distribution distribution of the data given $\{A(\vec{k},\omega_i)\}$ (the *likelihood* function [9]) has the form

$$P\left[G(\vec{k},\tau)|A(\vec{k},\omega)\right] \propto e^{-\chi^2/2} , \tag{8}$$

with

$$\chi^2 = \sum_i \frac{\left[G(\vec{k},\tau_i) - \bar{G}(\vec{k},\tau_i)\right]^2}{\sigma_i^2} . \tag{9}$$

By maximizing eq. (8) (or minimizing χ^2) a least squares fit to the data would be obtained.

In the context of maximum-entropy, a further probability distribution is introduced, that gives the probability distribution of the different spectral functions in the absence of data (*prior probability* [9]):

$$P\left[A(\vec{k},\omega)\right] \propto e^{\alpha S} \quad , \tag{10}$$

where S is the information theory entropy of A, with the following form [9]

$$S = \sum_i \Delta\omega \left[A(\omega_i) - m(\omega_i) - A(\omega_i)\ln\left(\frac{A(\omega_i)}{m(\omega_i)}\right)\right] \; . \tag{11}$$

Here is $m(\omega)$ the *default model* since it determines $A(\omega)$ in the absence of data. It general it is chosen such that it is consistent with prior knowledge about $A(\omega)$ (e.g. sum rules). The parameter α is adjusted self-consistently to a value $\hat{\alpha}$ corresponding to the maximum of $P\left[\alpha|G,m\right]$, a procedure that characterizes the *classical* maximum-entropy [9].

The spectral function is finally determined by maximizing the probability of $A(\omega)$ given the data $G(\tau)$. This *posterior* probability is given by the product of the likelihood function and the prior probability (Bayes' theorem [9]):

$$P\left[A(\omega)|G(\tau)\right] \propto P\left[G(\tau)|A(\omega)\right] \times P\left[A(\omega)\right] \; , \tag{12}$$

that amounts to a maximization of the exponent

$$\frac{\partial}{\partial A}\left(\hat{\alpha}S - \frac{\chi^2}{2}\right)\bigg|_{\hat{A}} = 0 \; , \tag{13}$$

where $\hat{A}$ is the optimal *image*. Maximum-entropy allows also an estimation of errors. We refer to the original literature on this point [9]. Further technical details can also be found in previous applications of the maximum-entropy method to the single impurity Anderson model [11], the one- [12] and two- [13] dimensional Heisenberg antiferromagnet.

In the present case, the only prior knowledge used was the positivity of the spectral functions. We have chosen an uninformative default model $m_i = \varepsilon$, where ε is a small quantity that merely suppresses noise in regions of insufficient information. The regularization parameter α has been determined selfconsistently by classical maximum-entropy method.

3 Results

We first compare the density of states $D(\omega)$ of a ring with 12 sites between data from QMC simulations and exact diagonalization (ED), both for half-filling (Fig. 1a) and for the doped case (Fig. 1b) with $< n >= 0.833 \simeq 5/6$. Although the maximum-entropy method is not able to resolve the rich structure obtained in ED, there is a good agreement in the shape and distribution of weight in the half-filled case. In the doped case, the maximum-entropy method can reproduce well the structures close to the Fermi-energy but has the general tendency of shifting weight to lower energies as one goes to higher energies due to the Laplace-transformation kernel. In particular, the pseudogap observed in ED between those states, stemming from the upper Hubbard band (UHB) brought by doping down to the top of the lower Hubbard band (LHB) [35], and the remaining of the UHB, does not appear in the maximum-entropy data.

In the following the data obtained for our largest system ($N = 84$), well beyond the capability of ED are shown [36]. We discuss first the spectral properties at half-filling.

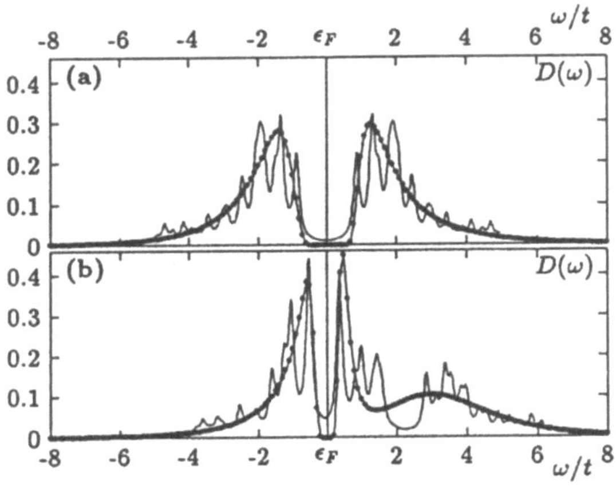

Figure 1: A comparison between QMC/MEM (lines with dots) and ED (full line) of the density of states for $N = 12$ sites: (a) Half-filling; (b) $< n >= 5/6$ ($\beta = 16$, $U = 4$).

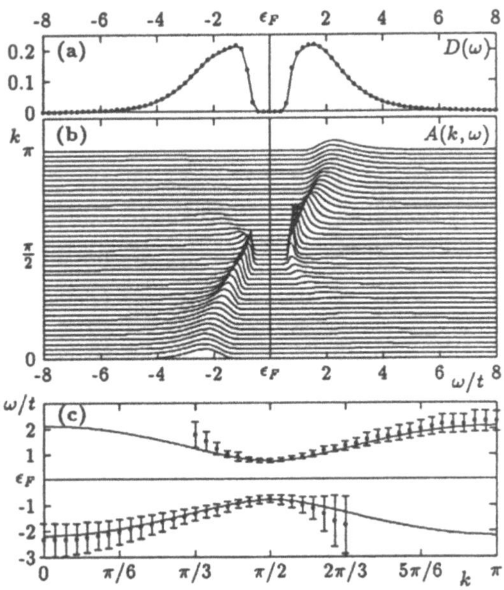

Figure 2: One-particle excitation for 84 sites at half-filling: (a) Density of states; (b) $A(k,\omega)$; (c) maxima of $A(k,\omega)$, the solid line is a cosine function adjusted to the bandwidth. ($\beta = 20$, $U = 4$)

Figure 2a shows $D(\omega)$ and the spectral function $A(k,\omega)$ is shown in Fig. 2b, where a dispersion can be clearly seen. Figure 2c shows the location of the maxima of $A(k,\omega)$ with the errors assigned by the MEM. These errors give the uncertainty in the location of those maxima. The full curve is the result of a SBMF calculation, where antiferromagnetic order is assumed. A remarkable coincidence is obtained. However, the spectral weight is shared quite differently, since although most of the weight of the LHB appears for $k \lesssim \frac{\pi}{2}$,

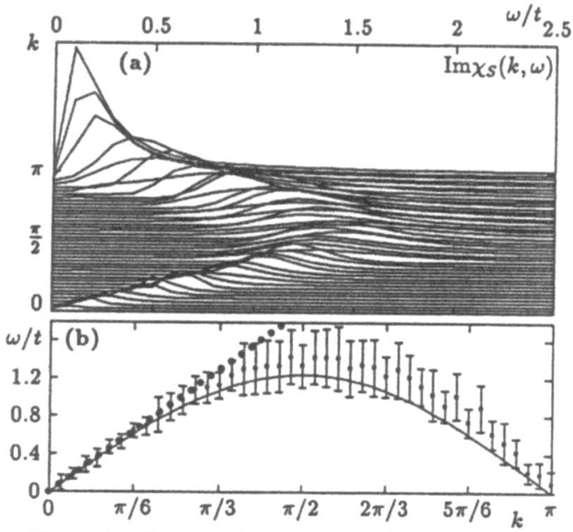

Figure 3: Spin-excitations at half-filling: (a) $\mathrm{Im}\chi_S(k,\omega)$; (b) maxima of $\mathrm{Im}\chi_S(k,\omega)$, the dotted line gives the slope at $k = 0$ ($v_s/t = 1.23 \pm 0.11$), and the full line stems from Bethe-Ansatz.

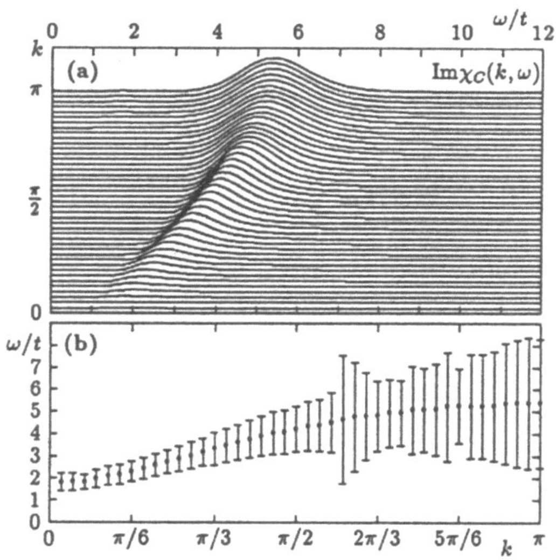

Figure 4: Charge-excitations at half-filling: (a) $\mathrm{Im}\chi_C(k,\omega)$; (b) maxima of $\mathrm{Im}\chi_C(k,\omega)$.

and the weight of the UHB is mostly concentrated in the region $k \gtrsim \frac{\pi}{2}$, around $k = \frac{\pi}{2}$ weight is splitted between the two bands such that as k goes from values below to values above k_F, spectral weight is continuously transferred from the lower to the upper Hubbard band.

Figure 3 shows the imaginary part of the spin-susceptibility $\chi_S(k,\omega)$ and the dispersion extracted from the maxima of Fig. 3a. From the slope of the dispersion around $k = 0$, we calculate a spinon-velocity $v_s/t = 1.23 \pm 0.11$. The exact value extracted from BA [18] $v_s^{BA} = 2tI_1(2\pi/U)/I_0(2\pi/U)$, where I_0 and I_1 are modified Bessel functions, is for $U/t = 4$, $v_s^{BA}/t = 1.2263$ [37], that compares remarkably well with our value. The full line in Fig. 3b corresponds to the dispersion extracted from Bethe Ansatz [18] showing that the agreement with the exact results in the thermodynamic limit is not limited to $\omega \to \infty$ but extends to finite frequencies. The equivalent quantities to those in Fig. 3 but now for charge-density excitations, are displayed in Fig. 4. We observe a rather broad structure with most of the weight around U, with a gap for $k \to 0$ that agrees within the errors with the one obtained in the one-particle spectrum.

Now we consider the doped case, where for definiteness we have chosen a hole doping $\delta \simeq \frac{1}{6}$. Figure 5 shows $D(\omega)$, $A(k,\omega)$, and the dispersion of the maxima in $A(k,\omega)$. A depression in $D(\omega)$ is observed at the Fermi-energy ϵ_F, that we assign to the fact that a Luttinger liquid has a density of states $D(\omega) \sim\mid \omega \mid^\alpha$, with $\alpha \simeq 0.038$ for $U/t = 4$ [38]. The vanishing of $D(\omega)$ at the Fermi energy cannot be seen in the simulation, since due to the small power α, an unrealistically high resolution would be necessary. Again, the "band-structure" obtained from $A(k,\omega)$ can be very well described by the SBMF paramagnetic band for $U = 4t$ within the error bars. At U=4 we still do not observe significant weight in $A(k,\omega)$ due to the UHB. However as was discussed in connection with ED, this can be due to the shifting of weight to lower frequencies by the MEM.

The spectra for spin-excitations are shown in Fig. 6. Again, a linear dispersion appears for low energies around $k = 0$ (inset of Fig. 6). Unfortunately the same resolution is not obtained around $2k_F = 5\pi/6$. However, these data can be combined with the corresponding ones for charge-density excitations (Fig. 7). With the values of spinon and holon velocities extracted from the dispersion curves, it is possible to obtain the conformal charge of the Hubbard model, that is known exactly to be $c = 1$. Such a relationship

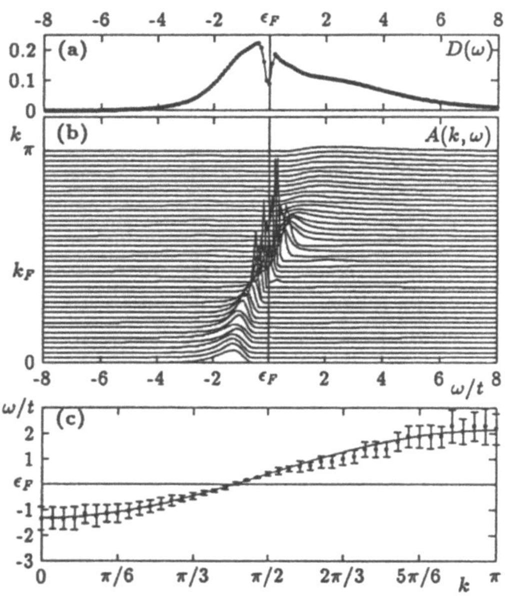

Figure 5: The same as Fig. 2, but now for $< n >= 5/6$.

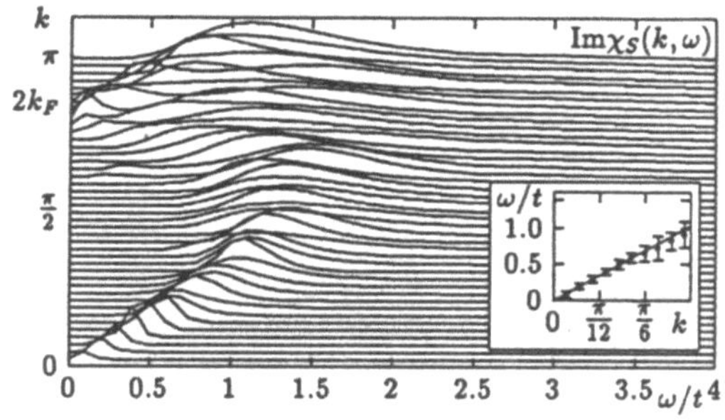

Figure 6: $\mathrm{Im}\chi_S(k,\omega)$ for $<n>= 5/6$. The inset shows the linear part of the dispersion around k=0. The slope gives $v_s/t = 1.33 \pm 0.09$.

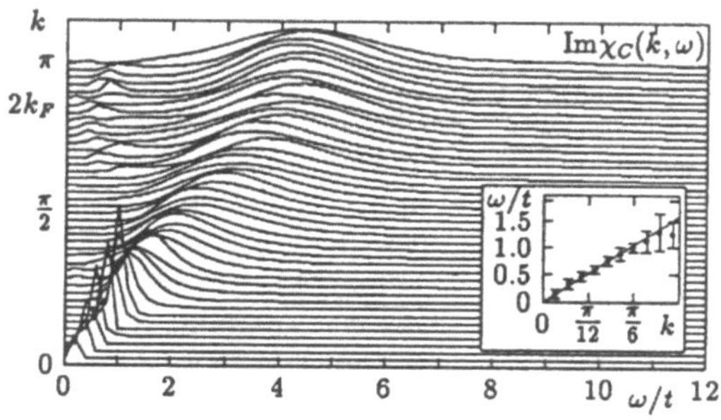

Figure 7: The same as Fig. 6 but for the charge susceptibility ($v_c/t = 2.02 \pm 0.14$).

stems from the conformal invariance of the model and has the following form [23]:

$$\frac{E_0(N)}{N} - \epsilon_0 = -\frac{\pi}{6N^2}(v_s + v_c)c + \mathcal{O}\left(\frac{1}{N^2(\ln N)^3}\right), \tag{14}$$

where $E_0(N)$ is the ground-state energy of the system with N sites, ϵ_0 is the ground-state energy per site in the thermodynamic limit, v_s and v_c are the spinon and holon velocities, respectively, and c is the conformal charge of the model. The value obtained for $N = 84$ is $c = 0.98 \pm 0.05$, where E_0 and ϵ_0 were determined from Bethe-Ansatz [17]. Such a stringent test shows that QMC together with the MEM are able to give reliable real frequency data for low-lying excitations of the Hubbard model. It should be stressed, that the insets in Figs. 6 and 7 make manifest charge-spin separation in the model, since $v_c > v_s$. This is clearly seen in all system sizes simulated.

Finally, we consider an exact sum-rule for the first frequency-moment of $\mathrm{Im}\chi_S(k,\omega)$ [28]:

$$\begin{aligned}\int_{-\infty}^{\infty} \frac{d\omega}{2\pi}\, \omega \,\mathrm{Im}\chi_S(k,\omega) &= -<[[H, S^z(k)], S^z(-k)]> \\ &= -2t(1-\cos k) < H_{kin} >, \end{aligned} \tag{15}$$

where $S^z = \frac{1}{2}(n_\uparrow - n_\downarrow)$ and $< H_{kin} >$ is the expectation value of the kinetic energy, that can be accurately calculated by QMC. The same sum-rule holds for $\mathrm{Im}\chi_C(k,\omega)$ but with S^z replaced by $n = n_\uparrow + n_\downarrow$. Figure 8 shows that the sum rule is fulfilled accurately over

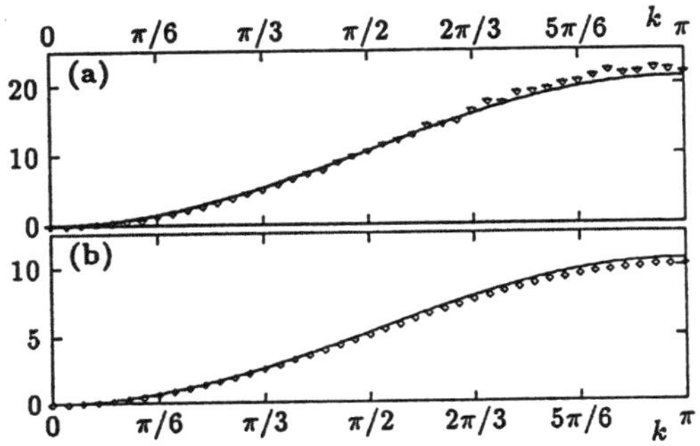

Figure 8: Sum-rule for the first frequency-moment ($< n >= 5/6$): (a) $\mathrm{Im}\chi_S(k,\omega)$; (b) $\mathrm{Im}\chi_C(k,\omega)$. The solid line shows the exact result.

most of the Brillouin-zone both for the spin- (Fig. 8a) and charge-susceptibilities (Fig. 8b) (deviations of less than 1%), with the exception of $k \gtrsim 2k_F$, where deviations ($\sim 10\%$) are obtained. They are probably due to a broad continuum similar to the one present in the 1-D Heisenberg antiferromagnet around $k = \pi$ [39, 40]. This result together with the conformal charge obtained, demonstrate the degree of reliability of the numerical data from low to intermediate frequencies and for very large systems that are well beyond the capability of other methods like ED.

Summarizing, we have presented real-frequency spectra for one-particle, spin- and charge-density excitations in the 1-D Hubbard model both in the insulating and in the metallic phases. The dispersion of the one-particle excitations is obtained for the first time. It agrees remarkably well with SBMF bands both at half-filling and in the doped case in spite of the non-Fermi-liquid character of the system. The dispersion obtained for the spin-excitations at half-filling agrees with results from BA within the error bars. In the doped case, charge-spin separation is obtained, where the spinon and holon velocities lead to an excellent agreement with the exact value for the conformal charge of the Hubbard model. The quality of the spectral data at intermediate frequencies is checked by an exact sum-rule, showing that only around $2k_F$, a departure of around 10% is present. The comparison with such a broad spectrum of exact results demonstrates the reliability of QMC in conjunction with the MEM for the HM.

Acknowledgements

We would like to thank T. Pruschke for instructive discussions. We are grateful to the Bavarian "FORSUPRA" program on high T_c research for financial support. The calculations were performed at the Cray YMP of the HLRZ Jülich and at LRZ München under a cooperation program with Cray Research Inc. We thank the above institutions for their support. This research was supported in part by the National Science Foundation under Grant No. PHY89-04035 and in part by the Deutsche Forschungsgemeinschaft under Grant No. Mu 820/7-1. AM is grateful to the Institute for Theoretical Physics, University of California, Santa Barbara, where this work was completed for its hospitality.

References

[1] E. Dagotto, to appear in Review of Modern Physics and references therein.

[2] D.J. Scalapino, UCSB preprint 1993, and references therein.

[3] S. Sorella et al., Europhys. Lett. **8**, 663 (1989).

[4] N. Furukawa and M. Imada, J. Phys. Soc. Japan 61, 3331 (1992).

[5] H.-B. Schüttler and D.J. Scalapino, Phys. Rev. Lett. **55**, 1204 (1985); Phys. Rev. B**34**, 4744 (1986).

[6] S.R. White, D.J. Scalapino, R.L. Sugar, and N.E. Bickers, Phys. Rev. Lett. **63**, 1523 (1989).

[7] M. Jarrell and O. Biham, Phys. Rev. Lett. **63**, 2504 (1989).

[8] R.N. Silver, D.S. Sivia, and J.E. Gubernatis, Phys. Rev. B**41**, 2380 (1990).

[9] J.E. Gubernatis, M. Jarrell, R.N. Silver, and D.S. Sivia, Phys. Rev. B**44**, 6011 (1991).

[10] S.R. White, Phys. Rev. B**44**, 4670 (1991); Phys. Rev. B**46**, 5678 (1992).

[11] R.N. Silver, et al., Phys. Rev. Lett. **65**, 496 (1990).

[12] J. Deisz, M. Jarrell, and D. Cox, Phys. Rev. B**42**, 4869 (1990).

[13] M. Makivić and M. Jarrell, Phys. Rev. Lett. **68**, 1770 (1992).

[14] N. Bulut, D.J. Scalapino, and S.R. White, Phys. Rev. Lett. **72**, 705 (1994).

[15] P.W. Anderson, Phys. Rev. Lett. **64**, 1839 (1990); **65**, 2306 (1990); Science **256**, 1526 (1992).

[16] F.D.M. Haldane, J. Phys. C **14**, 2585 (1981).

[17] E.H. Lieb and F.Y. Wu, Phys. Rev. Lett. **20**, 1445 (1968).

[18] M. Takahashi, Prog. Theor. Phys. **43**, 1619 (1970).

[19] N. Kawakami and A. Okiji, Phys. Rev. B **40**, 7066 (1989).

[20] H.J. Schulz, Phys. Rev. Lett. **64**, 2831 (1990).

[21] B.S. Shastry and B. Sutherland, Phys. Rev. Lett. **65**, 243 (1990); N. Kawakami and S.-K. Yang, Phys. Rev. Lett. **65**, 3063 (1990).

[22] S. Sorella and A. Parola, J. Phys.: Cond. Matter **4**, 3589 (1992).

[23] H. Frahm and V.E. Korepin, Phys. Rev. B **42**, 10553 (1990).

[24] R. Preuss, A. Muramatsu, W. von der Linden, P. Dieterich, F. F. Assaad and W. Hanke, to be published.

[25] G. Kotliar and A. Ruckenstein, Phys. Rev. Lett. **57**, 1362 (1986).

[26] W.E. Pickett, Rev. of Mod. Phys. **61**, 433, (1989).

[27] R. Lui et al., Phys. Rev. B**45**, 5614, (1992).

[28] P.C. Hohenberg and W.F. Brinkman, Phys. Rev. B**10**, 128 (1974).

[29] S.R. White, et al., Phys. Rev. B**40**, 506 (1989).

[30] R. Blankenbecler, D.J. Scalapino, and R.L. Sugar, Phys. Rev. D**24**, 2278 (1981).

[31] J.E. Hirsch, Phys. Rev. B**31**, 4403 (1985).

[32] E. Loh and J. Gubernatis, in Electronic Phase Transitions, ed. by W. Hanke and Y.V. Kopaev (North Holland, Amsterdam, 1992).

[33] J.E. Hirsch, Phys. Rev. B**28**, 4059 (1983).

[34] E.Y. Loh, J.E. Gubernatis, R.T. Scalettar, S.R. White, D.J. Scalapino, and R.L. Sugar. Phys. Rev. B**41**, 9301 (1990).

[35] H. Eskes, M.B. Meinders, and G.A. Sawatzky, Phys. Rev. Lett. **67**, 1035 (1991); M.B. Meinders, H. Eskes,and G.A. Sawatzky, Phys. Rev. B**48**, 3916 (1993).

[36] Results for smaller systems and further figures will be presented elsewhere.

[37] The BA results refer to the thermodynamic limit.

[38] H.J. Schulz, Int. J. Mod. Phys. B**5**, 57 (1991).

[39] J.M.P. Carmelo and P. Horsch, Phys. Rev. Lett. **68**, 2831 (1992); J.M.P. Carmelo, P. Horsch, and A.A. Ovchinnikov, Phys. Rev. B**46**, 14728 (1992).

[40] G. Müller, H. Beck, and J.C. Bonner, Phys. Rev. Lett. **43**, 75 (1979).

Simulating $2d$ Fermions in Strong Magnetic Fields

G. Ortiz

Department of Physics, University of Illinois at Urbana-Champaign, Urbana, IL 61801, USA

Abstract. The *Fixed-Phase* method, a stochastic approach to deal with complex hermitian Hamiltonians, is reviewed in the context of fermions coupled to external electromagnetic sources. The method uses as a key ingredient a trial phase that plays the role of gauge function in the unitary transformation that maps the original fermion problem to a boson problem for the modulus of the wavefunction. In particular, we investigate the ground state of an ideal $2d$ electron gas in high magnetic fields at various densities and for filling fractions $\nu = 1/m$. At high electron densities, the Quantum Hall liquid is the stable phase despite the character of particle interactions. Magnetophonon correlations turn out to be essential to explain the transition to an electron Wigner crystal.

1. Introduction

The Fractional Quantum Hall Effect (FQHE) is one of the most striking many-body phenomena discovered in the last decade [1]. It concerns the transport properties of electrons confined in $2d$ and subjected to strong magnetic fields. Electron interactions are important to resolve the massive degeneracy of the non-interacting system and stabilize a gap to charge excitations. These incompressible ground states are new strongly-correlated states of matter.

On the other hand, stochastic methods to study many-fermion systems proved to be quite powerful in the description of correlation effects. Thus, we could in principle apply these techniques to learn about the nature of the ground state and low-energy physics of the FQHE. However, random-walk methods (with the exception of Variational Monte Carlo (VMC)) are restricted to treat systems with time-reversal symmetry, and the problem of electrons in external magnetic fields does break this symmetry. We have recently developed [2] a stochastic approach to deal with complex hermitian Hamiltonians where time-reversal is broken explicitly (*Fixed-Phase* (FP) method).

In this article, we are going to review the FP method in detail and apply it to the physics of the FQHE. Of particular importance is the determination of the phase boundary between a Laughlin liquid (LL) and a Wigner crystal (WC) as a function of the particle density and Landau-level filling factor ν. A transition has been suggested based upon recent experimental observations of a reentrant insulating-FQHE-insulating behavior around ν=1/5 and ν=1/3 for electron and hole systems respectively [3].

Springer Proceedings in Physics, Vol. 78
Computer Simulation Studies in Condensed-Matter Physics VII
Editors: D.P. Landau · K.K. Mon · H.-B. Schüttler

2. Stochastic Method for Systems with no Time-Reversal Symmetry

Stochastic methods which use branched random-walks employ a distribution function P as a probability measure to sample points in configuration space. Throughout this work the latter is considered as a Cartesian manifold of dimension dN, where d is the spatial dimensionality and a point in this space is denoted by $\mathcal{R} = (\mathbf{r}_1, \ldots, \mathbf{r}_i, \ldots, \mathbf{r}_N)$, where $\mathbf{r}_i$ and N are the coordinates and number of particles, respectively. If the system Hamiltonian is invariant under time-reversal and the energy eigenstate is non-degenerate, it is always possible for its wavefunction to be real [4]. However, many interesting physical phenomena (*i.e.*, particles in an external magnetic field or in a rotating frame) involve explicit breaking of the time-reversal invariance, which means that the energy eigenfunction must be complex. All random-walk implementations to date (with the exception of VMC) have required the wavefunction to be real in order to use sampling methods, and consequently complex Hamiltonian operators were not investigated.

In this Section we introduce a general stochastic method [2] to study problems where time-reversal invariance is broken explicitly. We are mainly concerned with the study of the ground state (zero temperature) properties of a quantum system [5]. Though the method turns out to be of broad applicability, we will illustrate it by the study of interacting spin-$\frac{1}{2}$ fermions coupled to external electromagnetic sources.

Let us start by writing the nonrelativistic Hamiltonian which governs the dynamics of our N fermion system in presence of an external electromagnetic potential $A_\mu = (\mathbf{A}(\mathbf{r}), \phi(\mathbf{r}))$.

$$\widehat{\mathbb{H}} = \sum_{i=1}^{N} \left[\frac{\mathbf{\Pi}_i^2}{2m^*} + e\ \phi(\mathbf{r}_i) \right] + \sum_{1 \leq i < j \leq N} v(|\mathbf{r}_i - \mathbf{r}_j|) + \Lambda(\mathcal{R}) + \widehat{\mathbb{H}}_S \ , \qquad (1)$$

where $\mathbf{\Pi}_i = \mathbf{p}_i + \frac{e}{c}\mathbf{A}(\mathbf{r}_i)$ is the kinetic momentum. The vector potential $\mathbf{A}$, whose curl is the external magnetic field $\mathbf{B}$, is given in the Coulomb gauge $(\nabla \cdot \mathbf{A} = 0)$ and ϕ is the scalar potential. We will always be concerned with general interacting potentials v which are finite almost everywhere and can be only singular at coincident points $(\mathbf{r}_i = \mathbf{r}_j)$. Finally, $\Lambda(\mathcal{R})$ represents an additional "external" potential (*i.e.*, the ion potential) and $\widehat{\mathbb{H}}_S$ is the Zeeman energy term which takes into account the coupling of the $\mathbf{B}$-field to the total spin S of the system. Hence, we are dealing with N fermions of mass m^* and charge $-e$ coupled, in principle, to both orbital and spin degrees of freedom.

In usual space-spin formalism, it is known that the complete N-fermion state $\Psi(X)$ belongs to $\mathcal{L}_N^2$ (the space of states of the system) defined as

$$\mathcal{L}_N^2 = \left\{ \Psi \mid \widehat{P}_{ij}\Psi = -\Psi \ , \text{ and } \|\Psi\| = \left[\int \mid \Psi(X) \mid^2 d\,X \right]^{1/2} < \infty \right\} \ , \quad (2)$$

where $X = (\mathcal{R}, \Sigma)$ ($\Sigma = (\sigma_1, \ldots, \sigma_i, \ldots, \sigma_N)$ are discrete spin variables) and $\widehat{P}_{ij}$ represents the permutation of the pairs $(\mathbf{r}_i, \sigma_i)$ and $(\mathbf{r}_j, \sigma_j)$. Since the system Hamiltonian can be written as $\widehat{\mathbb{H}} = \widehat{\mathbb{H}}_{\mathcal{R}}(\mathcal{R}) + \widehat{\mathbb{H}}_S(\Sigma)$, the many-body wavefunction $\Psi(\mathcal{R},\Sigma)$ can be written as a product of a coordinate and a spin function (or a linear combination of such products),

$$\Psi(\mathcal{R},\Sigma) = \Phi(\mathcal{R}) \otimes \chi(\Sigma) . \tag{3}$$

We want to construct N-fermion eigenstates of $\widehat{\mathbb{H}}$, Ψ , that are also eigenfunctions of the total spin S^2,

$$S^2 \, \Psi(X) = \hbar^2 \, s(s+1) \, \Psi(X) , \tag{4}$$

and this is always possible because $\left[\widehat{\mathbb{H}}, S^2\right] = 0$. Thus, the configuration part $\Phi(\mathcal{R})$ must have the right symmetry in order to account for the Pauli principle. It turns out that a coordinate state $\Phi(\mathbf{r}_1, \cdots, \mathbf{r}_k, \mathbf{r}_{k+1}, \cdots, \mathbf{r}_N)$ which is symmetrized according to the Young scheme [6] and has total spin $s = \frac{N}{2} - k$ will be antisymmetric in the variables $\mathbf{r}_1, \cdots, \mathbf{r}_k$, and antisymmetric in the variables $\mathbf{r}_{k+1}, \cdots, \mathbf{r}_N$. Moreover, Φ possesses the property of Fock's cyclic symmetry,

$$\left(\mathbb{1} - \sum_{j=k+1}^{N} \widehat{P}_{kj} \right) \Phi = 0 , \tag{5}$$

where, in this case, $\widehat{P}_{kj}$ refers to the transposition of particle coordinates $\mathbf{r}_k$ and $\mathbf{r}_j$. This last condition is a very useful one for testing the symmetry of a given coordinate function.

For a given total spin s we are thus left with the task of solving an stationary many-body Schrödinger equation $\widehat{\mathbb{H}}_{\mathcal{R}}\Phi(\mathcal{R}) = E\Phi(\mathcal{R})$, where $\Phi(\mathcal{R})$ satisfies the symmetry constraint discussed above. Let us reformulate this eigenvalue problem in terms of the modulus $|\Phi|$ and phase φ [7] of the wavefunction Φ, such that $\Phi(\mathcal{R}) = |\Phi(\mathcal{R})| \exp[i\varphi(\mathcal{R})]$.

Given the energy functional

$$E\left[|\Phi|;\varphi\right] = \frac{\left\langle \Phi \middle| \widehat{\mathbb{H}}_{\mathcal{R}} \, \Phi \right\rangle}{\langle \Phi \mid \Phi \rangle} , \tag{6}$$

we can consider this as a functional of Φ and Φ^* (its complex conjugate) or as a functional of $|\Phi|$ and φ. In the first, case performing independent variations of E on (Φ, Φ^*) the functions which render E stationary ($\delta E = 0$) satisfy the Schrödinger equation (and its complex conjugate). In the second case, the stationary condition on E reads ($\widehat{\mathbb{H}}_{\mathcal{R}} = \widehat{\mathbb{H}}^{\dagger}_{\mathcal{R}}$)

$$\begin{aligned} \Re e\left\{\exp\left[-i\varphi\right]\left(\widehat{\mathbb{H}}_{\mathcal{R}} - E\right)\Phi\right\} &= 0 \\ \Im m\left\{\exp\left[-i\varphi\right]\left(\widehat{\mathbb{H}}_{\mathcal{R}} - E\right)\Phi\right\} &= 0 \end{aligned} , \tag{7}$$

where $\Re e$ and $\Im m$ mean real and imaginary parts of the expressions in brackets, respectively. Eqs. (7) can be written in a more transparent way :

$$\widehat{H}\,|\Phi(\mathcal{R})| = \left[\sum_{i=1}^{N} \frac{\mathbf{p}_i^2}{2m^*} + V(\mathcal{R})\right] |\Phi(\mathcal{R})| = E\,|\Phi(\mathcal{R})| \tag{8}$$

$$\sum_{i=1}^{N} \nabla_i \cdot \left[|\Phi(\mathcal{R})|^2 \left(\nabla_i\, \varphi(\mathcal{R}) + \frac{e}{\hbar c}\mathbf{A}(\mathbf{r}_i)\right)\right] = 0 \ , \tag{9}$$

where the effective potential $V(\mathcal{R})$ is given by

$$V(\mathcal{R}) = \sum_{i=1}^{N} \left[\frac{\hbar^2}{2m^*} \left(\nabla_i\, \varphi(\mathcal{R}) + \frac{e}{\hbar c}\mathbf{A}(\mathbf{r}_i)\right)^2 + e\,\phi(\mathbf{r}_i)\right] + \sum_{1\leq i<j\leq N} v(|\mathbf{r}_i - \mathbf{r}_j|) + \Lambda \ . \tag{10}$$

Then, the original eigenvalue problem, $\widehat{\mathbb{H}}_{\mathcal{R}}\Phi(\mathcal{R}) = E\Phi(\mathcal{R})$, is equivalent to solve simultaneously two real multidimensional differential equations. One (Eq. (8)) results from variations of the modulus $|\Phi|$ and another (Eq. (9)) from independent variations of the phase φ [8].

In this way, we have mapped the original fermionic problem into a bosonic one for $|\Phi|$. An alternative way of looking at this result is to regard it as a gauge transformation of the original problem whose effect is to add a vector potential (gauge field), $\nabla_i\, \varphi(\mathcal{R})$, to the Hamiltonian, giving rise to a *fictitious* magnetic field. Note that the gauge field, which is a non-local function of the particle positions (*i.e.*, depends on the position of all the particles), contains information on the statistics of the particles.

So far we have made no approximations. In fact, we have just reformulated the original eigenvalue problem in terms of the modulus and phase of the wavefunction Φ. The essence of the *Fixed-Phase* (FP) method consists in making a choice for the phase, φ_T, and solving *exactly* the bosonic problem for $|\Phi|$ (Eq. (8)) using Quantum Monte Carlo methods. It is clear from our variational formulation that if we fix the phase φ, Eq. (8) is the one that renders the *Fixed-phase* energy functional stationary ($\delta E_{\varphi_T}[|\Phi|] = 0$); to prove that this is a minimum we proceed as in elementary quantum mechanics [9]. Then the FP method provides a variational upper bound for the exact ground state energy $\mathcal{E}_0$, $E_{\varphi_T}[|\Phi|] \geq \mathcal{E}_0$, and for a prescribed trial phase φ_T the *lowest* energy consistent with this phase.

For real symmetric Hamiltonian operators (time-reversal invariant), the FP method reduces to the *Fixed-Node* approximation [10] when we make the choice $\varphi_T = \pi\,[1 - \Theta(\Phi_T)]$, where Θ is the Heaviside unit step function. $\varphi_T(\mathcal{R})$ is a real function which can assume only two values: 0 or π (mod(2π)); it is the location of the discontinuities of φ_T that is important, *i.e.*, the nodal structure of Φ_T [11]. Now, the gauge field is purely of statistical nature and represents just a boundary condition. In this way Eq. (8) is solved in domains of the configuration space delimited by the nodes of the trial state Φ_T, where hard wall boundary conditions are imposed. Notice that Eq. (9) is exactly satisfied inside these domains if there is no **B**-field.

Once we have made the choice of phase φ_T, we have to solve a boson problem. In order to solve this problem we use projector methods (DMC) [10, 12] which are based on the property that in the asymptotic imaginary time (t) limit the Euclidean evolution operator acting on a parent state $|\Phi_T|$ projects out the ground state $|\Phi_0|$

$$|\Phi_0| \propto \lim_{t\to\infty} \exp\left[-t(\widehat{H} - E_T)\right] |\Phi_T| , \tag{11}$$

where E_T is a suitable trial energy which shifts the zero of the energy spectrum.

We can transform the time-dependent Schrödinger equation for $|\Phi|$ in Euclidean time t to a master equation for the *importance-sampled* distribution $P(\mathcal{R},t) = |\Phi_T(\mathcal{R})| \, |\Phi(\mathcal{R},t)|$ (in atomic units)

$$\frac{\partial P(\mathcal{R},t)}{\partial t} = \sum_{i=1}^{N} \nabla_i \cdot \left[\frac{1}{2}\nabla_i P(\mathcal{R},t) - \mathbf{F}_i(\mathcal{R})\, P(\mathcal{R},t)\right] - (E_L(\mathcal{R}) - E_T)\, P(\mathcal{R},t) , \tag{12}$$

where $\Phi_T(\mathcal{R}) = |\Phi_T(\mathcal{R})| \exp\left[i\varphi_T(\mathcal{R})\right]$ is a trial function used to *guide* the random-walk; $\mathbf{F}_i(\mathcal{R}) = \nabla_i \ln |\Phi_T|$ is the *drift velocity*; $E_L(\mathcal{R}) = |\Phi_T|^{-1}\, \widehat{H}\, |\Phi_T|$ is the local energy. At sufficiently long times $P(\mathcal{R}, t \to \infty) \to |\Phi_T(\mathcal{R})| \, |\Phi_0(\mathcal{R})|$ (up to a normalization constant), where $|\Phi_0|$ is the lowest energy state, compatible with the phase φ_T, which has a component in $|\Phi_T|$. In order to get this stationary distribution, E_T must be adjusted to be equal to the FP ground state energy E_0, given in turn by

$$E_0 = \lim_{t\to\infty} \langle E_L(\mathcal{R})\rangle_{P(\mathcal{R},t)} = \frac{\int d\mathcal{R}\; P(\mathcal{R}, t\to\infty)\, E_L(\mathcal{R})}{\int d\mathcal{R}\; P(\mathcal{R}, t\to\infty)} . \tag{13}$$

To generate the stochastic process that yields $P(\mathcal{R}, t \to \infty)$ as its asymptotic distribution, we use the integral representation of Eq. (12)

$$P(\mathcal{R}', t+\tau) = \int d\mathcal{R}\; G(\mathcal{R} \to \mathcal{R}', \tau)\, P(\mathcal{R}, t) , \tag{14}$$

with an *importance* Green's function given by

$$G(\mathcal{R} \to \mathcal{R}', \tau) = \frac{|\Phi_T(\mathcal{R}')|}{|\Phi_T(\mathcal{R})|} \,\langle \mathcal{R}' | \exp\left[-\tau(\widehat{H} - E_T)\right] | \mathcal{R}\rangle . \tag{15}$$

Eq. (12) is solved by iterating Eq. (14) many times (power method). The modulus of the trial state, $|\Phi_T(\mathcal{R})|$, affects the convergence and statistical fluctuations of the stochastic averages.

3. The Physics of the Quantum Hall Effect

There are many excellent reviews of the physics of the Quantum Hall Effect [1]. In this Section we just introduce the elements necessary to understand the main features of this fascinating phenomena. Then, the *Fixed-Phase* method we have developed in previous Section is applied to study the quantum transition between a Quantum Hall liquid and an electron Wigner crystal.

3.1. Background

The Quantum Hall Effect (QHE) was discovered in the context of semiconductor heterostructures subjected to strong **B**-fields (of the order of 20 Teslas) and at very low temperatures (T $\approx$ millikelvin). In order to understand the effect, the relevant dynamics of the layer of electrons that is trapped at the interface between the two semiconductors can be idealized as the one of a $2d$ interacting Fermi system. The QHE involves $2d$ electrons under the influence of strong perpendicular uniform magnetic fields and low T.

When the effect is observed, the Hall conductance (transverse component) is quantized and given by $\nu e^2/h$, where ν, the filling fraction, is a rational number. Simultaneously, the longitudinal component vanishes, meaning that the system is dissipationless. What is remarkable about the effect itself is the existence of plateau (*i.e.*, constant value of the Hall conductance over finite intervals of magnetic fields) at those values of ν. In fact, for a Galilean-invariant system, a simple classical argument [1] leads to those values of conductance but cannot explain the existence and stability of the plateau, which are precisely related to the presence of disorder and a gap to the charge excitation spectrum. This apparently simple behavior suggests that some fundamental physical principle underlies this phenomenon.

It is well-known that the nature of the energy spectrum of a free-electron system changes from a continuous to a discrete one when the magnetic field is turned on. Moreover, the discrete single-particle levels (called Landau levels) are massively degenerate, with one state per spin per flux quantum $\phi_0 = hc/e$, *i.e.*, a number of states per unit area equal to $1/2\pi\ell^2$, where $\ell = \sqrt{\hbar c/eB}$ is the magnetic length. When the particle densities correspond to fill integer number of Landau levels, ν is an integer and the many-body ground state of this non-interacting problem (plus weak disorder) allows us to explain the main features of the so-called Integer Quantum Hall Effect (IQHE). On the other hand, if ν is a fractional number the many-body ground state will be highly degenerate and it turns out that particle correlations play a dominant role in stabilizing it. A big step towards the understanding of the Fractional Quantum Hall Effect (FQHE) was made by Laughlin [13] whose analysis showed how the repulsive interactions between electrons stabilized an energy gap for filling fractions $\nu = 1/m$ ($\nu < 1$), with m an odd integer. Qualitatively, the Laughlin state is an incompressible quantum liquid and constitutes a fascinating new state of matter where the gap is entirely due to many-body effects.

In the following discussion, we are concerned with the study of the ground state properties of an ideal (no disorder) $2d$ quantum system, defined by the many-body Hamiltonian

$$\widehat{\mathbb{H}}_{\mathcal{R}} = \sum_{i=1}^{N} \frac{\mathbf{\Pi}_i^2}{2m^*} + \sum_{1\leq i<j\leq N} v(|\mathbf{r}_i - \mathbf{r}_j|) + \Lambda \ , \tag{16}$$

where Λ is the (homogeneous) background potential term which ensures the global charge neutrality of the system (neutralizing background). The vector potential $\mathbf{A}$ is given by $\mathbf{A}(\mathbf{r}) = (-By, 0)$ ($\mathbf{A}(\mathbf{r}) = (-By/2, Bx/2)$) in the Landau (symmetric) gauge. When we refer to particle interactions, v, we have in mind the $3d$-Coulomb potential ($v(r) = e^2/r$). At the end we are going to discuss briefly the effect of longer range interactions ($v(r) = -e^2 \ln(r)$).

There are three *independent* parameters characterizing the $2d$ electron gas in presence of a magnetic field at T = 0. They are the Wigner-Seitz radius $r_s = 1/\sqrt{\pi\,\rho}$, which measures interparticle spacing in units of the Bohr radius ($\rho = \rho_\uparrow + \rho_\downarrow$ is the areal density and $\rho_{\uparrow(\downarrow)}$ is the density of spin-up(down) electrons), the Landau level filling factor $\nu = 2\ell^2/r_s^2$, and the spin-polarization $\zeta = |\rho_\uparrow - \rho_\downarrow|/\rho$. Hereafter, we will assume that the spin degrees of freedom are frozen by the magnetic field and that no electric field is present, *i.e.*, we will consider the physical situation where Zeeman energies are so large that the electrons in the system are completely spin-polarized ($\zeta = 1$). It is the interplay between the two parameters (r_s, ν) that will determine the zero temperature phase diagram. As ν is decreased (larger $\mathbf{B}$-fields) or r_s is increased the $2d$ system is expected to undergo a transition from a Laughlin liquid (LL) to a Wigner crystal (WC) [14]. In the next Sections we are going to show how useful the FP method turns out to be in order to study this transition.

3.2. Special Set of Admissible Phases

Thus, we are left with the task of exploring the set of non-equivalent phases $\{\varphi_T\}$ and of determining the minimum of the functional $E_0\,[\varphi_T]$. Of course, we do not pretend to scan the space of all possible phases but just use some conventional choices for φ_T. There are some mathematical constraints that can be imposed on φ_T. For instance, we would like to conserve the symmetries of the Hamiltonian unless some of them are spontaneously broken (like in the WC phase). In any case, it is obvious that a phase that satisfies Eq. (9) (*continuity equation*) for the exact $|\Phi(\mathcal{R})|$ will lead to the exact solution of the many-fermion problem. Thus, trial functions Φ_T that satisfy the *continuity equation* are good candidates for the phase. Consider trial functions of the form

$$\Phi_T(z_1, z_2, \cdots, z_N) = F(\{\bar{z}_i\}) \exp\left[-\Omega(\{z_i, \bar{z}_i\})\right] \quad , \tag{17}$$

where $F(\{\bar{z}_i\})$ is an analytic function of $\{\bar{z}_i\}$ (*i.e.*, it belongs to the Bargmann space of holomorphic functions), and $\Omega(\{z_i, \bar{z}_i\}) = \sum_{j=1}^{N} |z_j|^2/4\ell^2$ in the symmetric gauge or $\Omega(\{z_i, \bar{z}_i\}) = \sum_{j=1}^{N} y_j^2/2\ell^2$ in the Landau gauge. Particle positions are written in complex coordinate representation $z_j = x_j + iy_j$ ($\bar{z}_j = x_j - iy_j$). It is straightforward to prove [15] using complex analysis that the above conditions on Φ_T imply

$$\nabla_i \cdot \left[|\Phi_T(\mathcal{R})|^2 \left(\nabla_i\, \varphi_T(\mathcal{R}) + \frac{e}{\hbar c}\mathbf{A}(\mathbf{r}_i)\right)\right] = 0 \quad , \tag{18}$$

$$\nabla_i \cdot \left(\nabla_i\, \varphi_T(\mathcal{R}) + \frac{e}{\hbar c}\mathbf{A}(\mathbf{r}_i)\right) = 0 \tag{19}$$

and

$$E_L(\mathcal{R}) \;=\; \frac{\hbar^2 N}{2m^*\ell^2} + \sum_{1\leq i<j\leq N} v(|\mathbf{r}_i - \mathbf{r}_j|) \,+\, \Lambda \,. \tag{20}$$

Note that Eq. (18) is just the desired second equation, Eq. (9), except that the exact Φ is replaced by the trial Φ_T. Thus for this class of functions, Eq. (9) is approximately satisfied analytically. Furthermore, the last equation, Eq. (20), shows that the local energy needed in the sampling is given by an extremely simple form. These results are independent of the manifold in which particles move, *i.e.*, of the boundary conditions.

In order to gain further insight, it is instructive to analyze the phase of non-interacting particles. For simplicity, assume that the system consists of only two electrons ($\zeta = 1$). Then, the system Hamiltonian in the symmetric gauge is given by [16]

$$\widehat{\mathbb{H}}_0 = \sum_{i=1}^{2} \frac{\Pi_i^2}{2m^*} = \frac{\hbar^2}{m^*}\sum_{i=1}^{2}\left[-2\left(\partial_{\bar{z}_i} - \frac{z_i}{4\ell^2}\right)\left(\partial_{z_i} + \frac{\bar{z}_i}{4\ell^2}\right) + \frac{1}{2\ell^2}\right] . \tag{21}$$

It is clear that the 2-fermion ground state is multiply degenerate. Suppose that we want to construct an energy eigenstate of total angular momentum $L_{\mathbf{z}}\Phi = 3\hbar\Phi$, where $L_{\mathbf{z}} = \hbar\sum_{i=1}^{2}(\bar{z}_i\partial_{\bar{z}_i} - z_i\partial_{z_i})$. Then, all antisymmetric (unnormalized) ground states can be written

$$\Phi_\alpha(z_1,z_2) = \left[\bar{z}_1^3 - \bar{z}_2^3 + \alpha\left(\bar{z}_2^2\bar{z}_1 - \bar{z}_2\bar{z}_1^2\right)\right]\,\exp\left[-\frac{1}{4\ell^2}\sum_{j=1}^{2}|z_j|^2\right] , \tag{22}$$

where α is an arbitrary parameter. Unless other symmetries resolve the degeneracy, the phase $\varphi = \Im m\,\ln\left[\bar{z}_1^3 - \bar{z}_2^3 + \alpha\left(\bar{z}_2^2\bar{z}_1 - \bar{z}_2\bar{z}_1^2\right)\right]$ is arbitrary. However, we can always choose a particular α and work within this sector (for instance, $\alpha = 3$ leads to $\varphi = 3\,\Im m\,\ln(\bar{z}_1 - \bar{z}_2)$).

3.3. Generalized Periodic Boundary Conditions

So far we have not discussed about boundary conditions. We are going to assume that the fermions move on a $2d$ torus, because we want to consider both crystal and liquid states on an equal footing.

It is well known that assigning boundary conditions to quantum systems in a closed manifold is a delicate matter. If the system is constrained to move on the surface of a flat torus and the **B**-field is zero (*i.e.*, periodic orbits along the principal directions $\mathbf{L_1}$ and $\mathbf{L_2}$), then we can consistently impose *generalized Bloch conditions* on the wavefunctions whenever we perform a uniform translation of the system by a lattice vector $\mathbf{L} = \mathrm{n}_1\mathbf{L_1} + \mathrm{n}_2\mathbf{L_2}$

$$\Phi_T(\{\mathbf{r}_j + \mathbf{L}\}) \;=\; \exp\left[\imath\,\mathbf{k}\cdot\mathbf{L}\right]\Phi_T(\{\mathbf{r}_j\}) \,, \tag{23}$$

since $\left[\widehat{\mathbb{H}}, \widehat{\mathbb{P}}\right] = 0$ ($\widehat{\mathbb{P}}$ is the total linear momentum). $\mathbf{L_1} \wedge \mathbf{L_2}$ defines the principal region of the torus, $\mathrm{n}_1, \mathrm{n}_2$ are arbitrary integers, and $\mathbf{k}$ is a twist angle vector. Actually, the fact that for a given observable, averaging or choosing a particular **k**-boundary condition yields the same result (in the thermodynamic limit), can be considered the condition for this observable to be a bulk property. In other words, in the thermodynamic limit a bulk observable is insensitive to the boundary conditions.

On the other hand, if a magnetic field is present consistent boundary conditions for the many-body state (Φ_T) and vector potential ($\mathbf{A}$) imply that both must change by a "large" gauge transformation in order to allow a non-zero quantized external flux (these are generalized periodic boundary conditions [17]). Hence, Φ_T must be a quasi-periodic function under a uniform translation of the system by a lattice vector $\mathbf{L}$

$$\Phi_T(\{\mathbf{r}_j + \mathbf{L}\}) = \exp\left[\imath\,\mathbf{k}\cdot\mathbf{L} + \imath\sum_{j=1}^{N}\Delta_j(\mathbf{L})\right]\Phi_T(\{\mathbf{r}_j\})\,. \tag{24}$$

$\Delta_j(\mathbf{L})$ is the gauge function given by $\Delta_j = x_j L_y/\ell^2$ ($\Delta_j = (x_j\,L_y - y_j\,L_x)/2\ell^2$) in the Landau (symmetric) gauge.

Note that even if toroidal boundary conditions lead to quasi-periodic states, the quantities in Eq. (12) ($|\Phi_T|$, $\mathbf{F}_i$) are genuine periodic functions on the torus. This fact simplifies the Monte Carlo sampling procedure.

3.4. Study of the 2*d* Solid-Liquid Transition

In this Section we are going to apply the FP method to study the transition between liquid and crystal states. To this end, we will consider electrons moving on the surface of a torus ($\mathbf{L_1} = (L_1, 0), \mathbf{L_2} = (0, L_2)$) whose ground state is completely spin-polarized and with boundary angle $\mathbf{k} = \mathbf{0}$. For the liquid we use the phase which results from Laughlin's wavefunction ($\nu = 1/m$), while for the crystal we consider two different phases (one includes magnetophonon correlations) which explicitly break rotational and translational invariance.

The unnormalized $\nu = 1/m$ Laughlin state, Φ_m, with toroidal boundary conditions has the form referred to in Eq. (17) with (Landau gauge) [18]

$$F(\{\bar{z}_i\}) = \exp\left[\imath k Z\right]\left[\vartheta_1\left(\frac{\pi(Z-Z_l)}{L_1}\middle|\,\tau\right)\right]^m \prod_{1\leq i<j\leq N}\left[\vartheta_1\left(\frac{\pi(\bar{z}_i-\bar{z}_j)}{L_1}\middle|\,\tau\right)\right]^m , \tag{25}$$

where ϑ_1 is the first odd elliptic theta function [19], $\tau = \imath\frac{L_2}{L_1}$, $Z = \sum_{j=1}^{N}\bar{z}_j$, $k = \frac{\pi N m}{L_1}$ and $Z_l = N\frac{L_1 - \imath L_2}{2}$ (all center-of-mass zeros at the same point). Due to the scaling properties of the Laughlin states with r_s, its kinetic and potential energies will satisfy the virial relation.

As a first check we use the Metropolis algorithm to compute the expectation value of the potential energy. The values are quoted in Table 1 where energies are expressed in units of e^2/ℓ. We have verified that, up to the sizes considered in our study ($N \sim 400$), finite-size effects are smaller than the statistical error and it turns out, when compared to the results of Ref. [20], that the energy is insensitive to boundary conditions.

Table 1: VMC energies ($E_0/N - \hbar\omega_c/2$) for the Laughlin wavefunction Φ_m in units of e^2/ℓ, for different filling factors ν. We have used generalized periodic boundary conditions. The numbers in parentheses represent the statistical error bar in the last decimal place.

ν=1/3	ν=1/5	ν=1/7
-0.4099(2)	-0.3277(1)	-0.2810(1)

Table 2: Fixed-Phase ground state energies ($E_0/N - \hbar\omega_c/2$) in units of e^2/ℓ for different electron densities r_s and filling factors ν. Columns display the energy of the quantum liquid state with Fixed-Phase φ_m. In all cases we have used generalized periodic boundary conditions.

ν	r_s			
	1	10	20	30
1/3	-0.4112(2)	-0.4150(2)	-0.4179(2)	-0.4201(2)
1/5	-0.32799(9)	-0.32902(9)	-0.32999(9)	-0.3308(1)
1/7	-0.28118(7)	-0.28164(8)	-0.28214(9)	-0.28262(9)

At this point, we start our *Fixed-Phase* computation with the Laughlin phase $\varphi_m = -\imath \ln[\Phi_m/|\Phi_m|]$. In order to generate the boson random-walk we followed the method described in Ref. [21] closely related to the one of Ref. [10]. The computation starts with an initial ensemble of N_c=200 walkers in configuration space ($\mathcal{R}_i, i = 1, \cdots, N_c$), distributed according to $P(\mathcal{R}, t = 0) = |\Phi_m|^2/||\Phi_m||^2$. To propagate the configurations of the ensemble in Euclidean time, we need to know the Green's function $G(\mathcal{R} \to \mathcal{R}', \tau)$. At each time step τ and for every configuration, we attempt to move each of the electrons by sampling the approximate Green's function (short-time approximation)

$$\widetilde{G}(\mathcal{R} \to \mathcal{R}', \tau) = \frac{1}{(2\pi\tau)^{3N/2}} \exp\left[-\frac{(\mathcal{R}' - \mathcal{R} - \tau\mathbf{F}(\mathcal{R}))^2}{2\tau}\right] \exp\left[-\tau\left(\frac{E_L(\mathcal{R}) + E_L(\mathcal{R}')}{2} - E_T\right)\right], \quad (26)$$

describing a diffusive random-walk of width $\sqrt{\tau}$ with a superimposed drift of velocity $\mathbf{F}$ and a branching process determined essentially by the second exponential term. The role of $\mathbf{F}$ is to guide the random-walk towards regions of phase space where $|\Phi_m|$ is larger. The total number of configurations is stabilized when E_T equals the mean value of the local energy (Eq. (13)), which in the large time limit equals the best upper bound to the true ground state energy. The results of this computation (E_0) are shown in Table 2 for N =45 spin-polarized electrons in a square Bravais lattice.

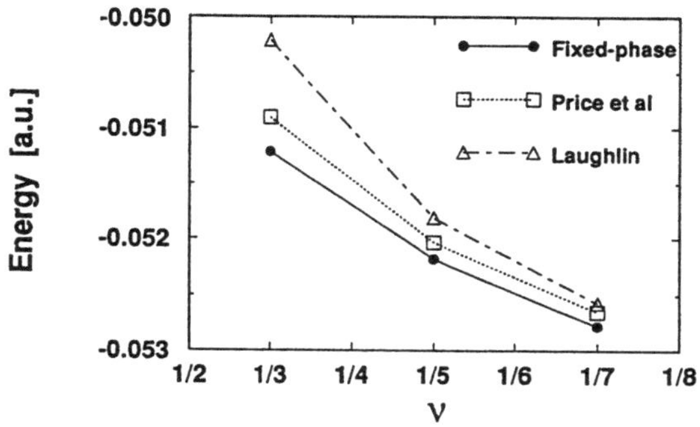

Figure 1: $(E_0/N - \hbar\omega_c/2)$ in effective atomic units for the Laughlin liquid at r_s=20. Laughlin stands for the VMC result using Laughlin wavefunction. The size of the symbols is proportional to the statistical uncertainty and lines are just a guide to the eye. Note that we have not drawn the cusps in the total energy at $\nu = 1/m$ (incompressibility).

In Fig. 1 we compare (in effective a.u.) the FP total energy to the Laughlin and Price *et al* recently proposed Jastrow-Slater wavefunction for r_s=20. It is clear from the figure that we get a substantial lowering of the liquid energies, which is relevant for studies of the liquid-solid transition.

Ground state expectation values of observables other than the total energy are, in general, more difficult to evaluate. For a general operator $\hat{\mathcal{O}}$ that does not commute with the Hamiltonian, the *mixed estimator* analogous to Eq. (13) can be written

$$\left\langle \hat{\mathcal{O}} \right\rangle^{mix} = \frac{\int d\mathcal{R}\, P(\mathcal{R}, t \to \infty)\, |\Phi_T|^{-1} \hat{\mathcal{O}} |\Phi_T|}{\int d\mathcal{R}\, P(\mathcal{R}, t \to \infty)} , \qquad (27)$$

where for the present purposes $\Phi_T = \Phi_m$. It is possible to have an expression that is accurate to second order in the difference $(|\Phi_0| - |\Phi_T|)$ by evaluating the *extrapolated estimator*:

$$\left\langle \hat{\mathcal{O}} \right\rangle^{ex} = 2 \left\langle \hat{\mathcal{O}} \right\rangle^{mix} - \frac{\left\langle \Phi_T \mid \hat{\mathcal{O}}\, \Phi_T \right\rangle}{\langle \Phi_T \mid \Phi_T \rangle} . \qquad (28)$$

Figure 2 shows the mixed estimator pair correlation functions defined as $g(r) = \frac{2}{\rho\,(N-1)} \left\langle \sum_i^N \sum_{j \neq i}^N \delta(r - r_{ij}) \right\rangle_{P(\mathcal{R}, t \to \infty)}$ for ν=1/7 at different r_s's. The physical picture that emerges is clear: as r_s increases (lower electron densities) more correlation is gained because of a more effective Landau level mixing.

Let us now consider a Wigner crystal state [22] using a phase which breaks explicitly the continuous translational and rotational invariance. To this end, we will consider the simplest trial state: a determinantal wavefunction

$$\Phi_{\mathrm{W}} = \hat{\mathcal{A}} \prod_{i=1}^{N} \phi_i(\mathbf{r}_i) \quad , \qquad (29)$$

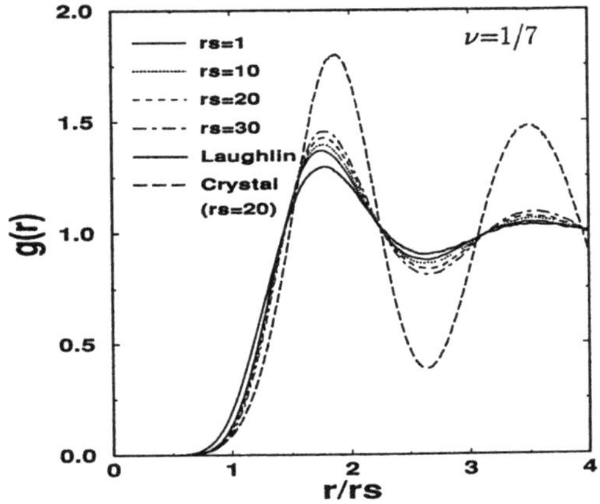

Figure 2: Mixed estimator pair correlation functions, $g(r)$, at ν=1/7 and various densities. We also display the VMC $g(r)$ that corresponds to the Laughlin wavefunction.

where $\widehat{\mathcal{A}}$ is the antisymmetrization operator and ϕ_i single-particle states given by (symmetric gauge) [23] $\phi_i = \frac{\beta}{\sqrt{2\pi\ell^2}}\ \widetilde{\phi}_i$, where

$$\widetilde{\phi}_i(\mathbf{r}) = \sum_{\mathbf{L}} \exp\left[-\frac{1}{4\ell^2}\left(\beta^2(\mathbf{r}-\mathbf{R}_i-\mathbf{L})^2 - 2\imath\left(\mathbf{r}\wedge\mathbf{R}_i + \mathbf{r}\wedge\mathbf{L} + \mathbf{R}_i\wedge\mathbf{L}\right)_z\right)\right], \tag{30}$$

with β a variational parameter which fixes the width of the orbital and $\mathbf{R}_i = r_s\sqrt{\frac{2\pi}{\sqrt{3}}}\left(i_1 + \frac{i_2}{2}; \frac{\sqrt{3}}{2}i_2\right)$ ($(i_1;i_2) \in \mathbb{Z}$) points of a regular $2d$ triangular lattice.

It turns out that, under the assumption of non-overlapping states $\langle\phi_i|\phi_j\rangle = \delta_{ij}$, the total energy $\left\langle \Phi_{\mathrm{W}} \middle| \widehat{\mathbb{H}}\Phi_{\mathrm{W}} \right\rangle_\beta$ can be evaluated analytically [15] and minimized with respect to β. The result of this calculation for r_s=20 is shown in Fig. 3 (HFLM), where we compare it to Lam and Girvin's magnetophonon wavefunction results (LG) [24].

We have assessed the importance of exchange in the total energy by using the determinantal wavefunction Φ_{W}. It is evident that in the low density regime particle exchanges do not affect the total energy (which is mainly Hartree) because of the rapid gaussian decay. In particular, we have checked that this is the case for the range of r_s studied. Thus, in order to apply the FP method, we have considered the phase that corresponds to the even simpler unsymmetrized Φ_{W} state, φ_W. Our results are described in Fig. 3 for $N = 56$ particles enclosed in a rectangular torus commensurate with a triangular lattice. Again, we have verified that finite-size effects are smaller than the statistical uncertainties. It is apparent from the figure that for this simple choice of phase, no transition to a WC is observed at $r_s = 20$ in the range of ν's considered.

It is clear that a state like Φ_{W} has no built-in correlations between particle-position fluctuations about lattice sites ($\mathbf{x}_i = \mathbf{r}_i - \mathbf{R}_i$). We can lower the

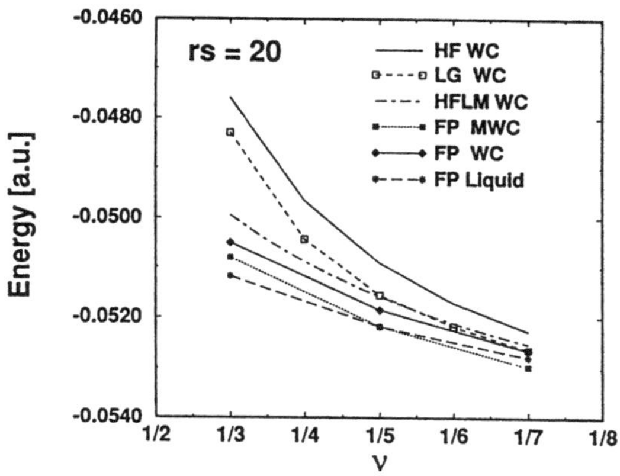

Figure 3: $(E_0/N - \hbar\omega_c/2)$ in effective atomic units for the Wigner crystal at r_s=20 resulting from different approximations. HF : Eq. (30) for β=1; LG : Lam and Girvin; HFLM : HF for optimized β; FP MWC : Fixed-Phase φ_{MP}, FP WC : Fixed-Phase φ_W. Comparison with our FP liquid energy (from Fig. 1) is also given.

potential energy of the system by correlating fluctuations at different sites. A way to accomplish that is by considering the ground state of the harmonic solid (magnetophonon Hamiltonian)

$$\widehat{\mathbb{H}}_{MP} = \sum_{i=1}^{N} \frac{\left[\mathbf{p}_i + \frac{e}{c}\mathbf{A}(\mathbf{x}_i)\right]^2}{2m^*} + \frac{1}{2} \sum_{i,j} x_\mu(i) D_{\mu,\nu}(i-j) x_\nu(j) \ , \qquad (31)$$

where $D_{\mu,\nu}$ is the dynamical matrix ($\mu(\nu)$ = 1,2). In the thermodynamic ($N, L \to \infty$) and large magnetic field [25] limits the distinguishable-particle ground state is given, in the symmetric gauge, by [24, 23, 15]

$$\Phi_{MP} = \prod_{i=1}^{N} \phi_i(\mathbf{r}_i) \ \exp\left[-\frac{1}{4\ell^2} \sum_{i,j} \xi_i \ B(i-j) \ \xi_j\right] \quad , \quad \xi_i = \bar{z}_i - \bar{R}_i \ , \qquad (32)$$

where the Fourier transform of $B(i-j)$ is

$$B(\mathbf{k}) = -\frac{\omega_L(\mathbf{k}) - \omega_T(\mathbf{k})}{\omega_L(\mathbf{k}) + \omega_T(\mathbf{k})} \exp\left[i\Theta_{\mathbf{k}}\right] \ , \qquad (33)$$

with

$$\exp\left[i\Theta_{\mathbf{k}}\right] = \frac{\delta}{\sqrt{\delta^2 + D_{12}^2}} + \frac{i \ D_{12}}{\sqrt{\delta^2 + D_{12}^2}} \quad , \quad \delta = \frac{D_{11} - D_{22}}{2} \ . \qquad (34)$$

Notice that for $\beta = 1$, Φ_{MP} has the form referred to in Eq. (17).

Eventually we want a wavefunction which displays magnetophonon correlations (like Φ_{MP}) and is defined on the torus. The construction of such a state

Table 3: Fixed-Phase ground state energies ($E_0/N - \hbar\omega_c/2$) in units of e^2/ℓ for different electron densities r_s and filling factors ν. Columns display the energy of the electron solid state (Wigner crystal) with Fixed-Phase φ_{MP}. In all cases we have used generalized periodic boundary conditions.

ν	r_s = 1	10	20	30
1/3	-0.3991(2)	-0.4106(3)	-0.4148(3)	-0.4183(3)
1/5	-0.3265(1)	-0.32868(9)	-0.3300(1)	-0.33118(9)
1/7	-0.2819(1)	-0.28255(9)	-0.28306(8)	-0.28359(8)

is not simple and, for the sake of simplicity, we prefer to adopt a pragmatic point of view: We assume a form like Eq. (32) with single-particle orbitals which satisfy the proper generalized boundary conditions (Eq. (30)). We can gauge the importance of this approximation by studying its size effects. In the thermodynamic limit the approximate state tends to Φ_{MP}, which is the exact magnetophonon state for strong **B**-fields. Precise technical details can be found elsewhere [15].

This state provides a phase φ_{MP} which we use as input for our FP method. Again, we dropped the antisymmetrization operator $\mathcal{A}$ because the exchange energy is negligible compared to correlation effects for the densities studied. The results for $N = 56$ electrons are displayed in Table 3. Statistical error bars are larger than finite-size effects. The energies we get for the solid phase ($r_s = 20$) are shown in Fig. 3 (FP MWC). We conclude that a transition to an electron WC is observed for ν=1/5 when compared to the best energies available for the Quantum Hall liquid (FP Liquid). Magnetophonon correlations in the phase φ_{MP} are essential to explain this transition.

So far, all the results pertain to electrons interacting via a $v(r) = e^2/r$ potential. It is usually asserted that the Laughlin liquid states are excellent approximations to the ground state of the $2d$ electron gas in strong **B**-fields regardless of the nature of particle interactions. In fact, Laughlin has shown [1], using exact diagonalization techniques, that for a three particle system the overlap between his ansatz and the exact ground state is larger than 99 % and insensitive to the range of the potential. We have investigated this issue [26] by considering the *true* $2d$ Coulomb liquid, *i.e.*, a $2d$ electron gas with logarithmic interactions. We find that, independent of the interactions, the Quantum Hall liquid is always the preferred state for low r_s and $\nu > 1/7$. The value of the critical r_s, *i.e.*, the density (r_s^c) for which we observe a liquid-solid transition, depends on ν. For instance, for $\nu = 1/5$, $r_s^c < 10$ (for the e^2/r potential it is of the order of 20). Thus, in a certain sense, the Laughlin state which is the RPA $r_s \to 0$ limit is adiabatically connected to the interacting state. In other words, the Laughlin ansatz is a stable fixed-point of the problem. The situation is comparable to what happens in Fermi liquid theory with the important

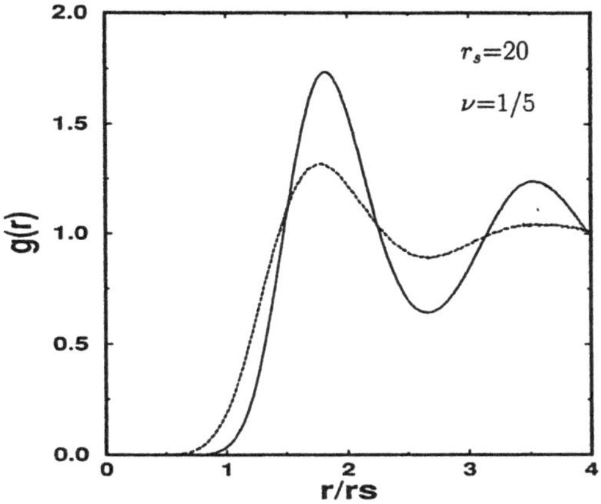

Figure 4: Mixed estimator pair correlation functions, $g(r)$, at r_s=20 and ν=1/5. Full line: corresponds to logarithmic particle interaction. Dash line: 3d-Coulomb interaction.

difference that the Laughlin states are incompressible. In Fig. 4 we compare the mixed estimator pair correlation functions for the same system but different potentials. We observe that for true long-range interactions ($v(r) = -e^2 \ln(r)$), the system is more crystal-like.

4. Summary and Conclusions

The *Fixed-Phase* method, a random-walk approach to studying problems where time-reversal invariance is broken explicitly, has been reviewed. The key ingredient is a trial phase taken from an approximate solution and playing the role of gauge function in the fermion-to-boson mapping of the original problem. Any choice for the phase provides a variational upper bound for the ground state energy of the system. This method is particularly suited to studying 2d interacting electron systems in presence of strong magnetic fields, *i.e.* the physics relevant to the Quantum Hall Effect problem. We applied this *Fixed-Phase* approach to investigate the transition at zero temperature between an incompressible $\nu = 1/m$ Laughlin liquid and a quantum Wigner crystal. For the electron solid we have used different phases and concluded that magnetophonon correlations are essential to explain the experimentally observed reentrant behavior at filling fraction $\nu = 1/5$ [3]. We also discussed the effect of long-range interactions ($v(r) = -e^2 \ln(r)$) on the stability of the Quantum Hall liquid and concluded that the Laughlin ansatz is a stable solution in the high-density regime (low r_s) for $\nu > 1/7$.

So far, we have considered the "simple" odd denominator filling fractions. Unfortunately, Laughlin's theory and its simplest extensions have nothing to say at filling fractions with even denominators. Of particular interest is the case

$\nu = 1/2$, because a Hall plateau has not been observed experimentally even in the highest quality samples, and because it is observed that it corresponds to a compressible fluid. It was recently proposed [27] that this state behaves like a system of electrons in zero average magnetic field. We think the FP method may help to clarify the nature of the $\nu = 1/2$ ground state.

Acknowledgments

I would like to acknowledge many useful discussions with David Ceperley, Richard Martin and Eduardo Fradkin. I am also indebted to Vivek Rao, who has proofread the manuscript with diligence and criticism. This work has been supported by NSF Grant No. DMR91-17822. The computations were done on the CRAY-YMP at the National Center for Supercomputing Applications.

References

[1] *The Quantum Hall Effect*, edited by R. E. Prange and S. M. Girvin (Springer-Verlag, New York, 1990), 2nd ed.; *Quantum Hall Effect*, edited by M. Stone (World Scientific, Singapore, 1992); E. Fradkin, *Field Theories of Condensed Matter Systems* (Addisson-Wesley, Redwood City, CA, 1991), Chapters 9,10; A. Karlhede, S. A. Kivelson and S. L. Sondhi, in *Correlated Electron Systems*, edited by V. J. Emery, (World Scientific, Singapore, 1993).

[2] G.Ortiz, D.M.Ceperley, and R.M.Martin, Phys. Rev. Lett. 71, 2777 (1993).

[3] H. W. Jiang, R. L. Willett, H. L. Stormer, D. C. Tsui, L. N. Pfeiffer and K. W. West, Phys. Rev. Lett. **65**, 633 (1990); M. B. Santos, Y. W. Suen, M Shayegan, Y. P. Li, L. W. Engel and D. C. Tsui, Phys. Rev. Lett. **68**, 1188 (1992).

[4] Elements of the Hilbert space $\mathcal{L}_N^2$ of a quantum mechanical system (*i.e*, wavefunctions) which differ *only* by a global phase are equivalent and said to belong to the same ray.

[5] It is possible to generalize the method to finite temperatures, allowing us to explore the thermodynamic properties of fermion systems with broken time-reversal symmetry (Ref. [15]).

[6] M. Hamermesh, *Group Theory* (Addison-Wesley, Reading, Massachusetts, 1962).

[7] All the phases φ are defined *modulo* 2π.

[8] Notice the formal similarity between Eq.(9) and the conservation of the "diamagnetic" component of the current density.

[9] A. Messiah, *Quantum Mechanics*, Vol. 2 (Interscience, New York, 1961).

[10] P. J. Reynolds, D. M. Ceperley, B. J. Alder, and W. A. Lester Jr., J. Chem. Phys. **77**, 5593 (1982); B. H. Wells, in *Methods in Computational Chemistry*, Vol. 1, edited by S. Wilson (Plenum, New York, 1987) p. 311.

[11] Coincident planes ($\mathbf{r}_i = \mathbf{r}_j$) are hyperplanes of dimension $dN - d$. If the only nodal surfaces were coincident planes we could solve the fermion problem *exactly*. In fact, this is what happens in $1d$; in higher dimensions the situation is more complicated. For a clear discussion of fermion nodes see D. M. Ceperley, J. Stat. Phys. **66**, 1237 (1991).

[12] D. M. Ceperley and M. H. Kalos, in *Monte Carlo Methods in Statistical Physics*, edited by K. Binder, (Springer Verlag, Berlin, 1979), p. 145; P. J. Reynolds, *et al.*, Int. J. Quant. Chem. **29**, 589 (1986).

[13] R. B. Laughlin, Phys. Rev. Lett. **50**, 1395 (1983).

[14] This is the experimental consensus, although there are probably other, more exotic, possibilities to explore.

[15] G. Ortiz, D. M. Ceperley and R. M. Martin (to be published).

[16] Note that our convention is for e and B to be positive. One must be careful in this respect, because some authors use different conventions.

[17] F. D. M. Haldane, Phys. Rev. Lett. **55**, 2095 (1985).

[18] F. D. M. Haldane and E. H. Rezayi, Phys. Rev. B **31**, 2529 (1985).

[19] I. S. Gradshteyn and I. M. Ryzhik, *Table of Integrals, Series and Products* (Academic, New York, 1980), p. 921.

[20] R. Price, P. M. Platzman, and S. He, Phys. Rev. Lett. **70**, 339 (1993).

[21] G. Ortiz and P. Ballone, Europh. Lett. **23**, 7 (1993); G. Ortiz and P. Ballone (preprint Urbana 1993).

[22] E. P. Wigner, Phys. Rev. **46**, 1002 (1934).

[23] X. Zhu and S. G. Louie, Phys. Rev. Lett. **70**, 335 (1993).

[24] P. K. Lam and S. M. Girvin, Phys. Rev. B **30**, 473 (1984); **31**, 613(E) (1985).

[25] In this context, large **B**-fields means that the cyclotron frequency ω_c is such that $\omega_c \gg \omega_L(\mathbf{k}) + \omega_T(\mathbf{k})$ for all **k**'s, where $\omega_{L(T)}$ are the longitudinal (transverse) vibrating modes of the $2d$ triangular lattice without a magnetic field.

[26] G. Ortiz and D. M. Ceperley (unpublished).

[27] B. I. Halperin, P. A. Lee and N. Read, Phys. Rev. **47**, 7312 (1993).

Part III

Computer Security

Computer Security in Scientific and Administrative Environments

R. W. Gerling

Max-Planck-Gesellschaft zur Förderung der Wissenschaften e.V., Generalverwaltung, Postfach 101 062, D-80084 München, Germany and Center for Simulational Physics, The University of Georgia, Athens, GA 30602, USA

Abstract. The use of and the communication via computer becomes widespread. Along with this more and more confidential data are stored in computers or send through computer networks. The unauthorized access to the confidential data has to be restricted. The standard to do so is the use of modern cryptographic tools. A short introduction to cryptography is given. The application of the tools to standard computer Security problems is demonstrated.

1. Introduction

Computers have become an essential part of our lives and as a consequence of that we relay on the data stored in our computers. From student grades to exams, from reports to research results, from referee reports to personal data lots of different kinds of sensitive data are stored in open computers (e.g. PC's and UNIX workstations) readily accessible by the unauthorized hacker. We are not talking about classified information here, only the sensitive data a standard university scientists has in his computer. Nobody would consider Security measures important; research is open and data sharing is standard. That is good and necessary for the advancement of science. Nevertheless some privacy is needed even in the most open environment. Also a lot of countries have dedicated data privacy laws, which require special car for the handling of personal data. A good overview gives Ref. 1.

2. Cryptography

Cryptography provides us with the tools to guarantee the privacy of our data. Over the years mathematicians have shown us how to use this tools and assured as about the quality. Cryptography is not comparable with alchemy and it also not a field of secret service agents and other relicts of the cold war. It makes heavy use of number theory and statistics and is a branch of modern mathematics. A good and easy to read introduction is the book by Beutelspacher [2]. Fumy and Rieß book [3] is a text book which gives a good theoretical background. More towards the programmer oriented is the book by Schneier [4].

Springer Proceedings in Physics, Vol. 78
Computer Simulation Studies in Condensed-Matter Physics VII
Editors: D.P. Landau · K.K. Mon · H.-B. Schüttler

2.1. Caesar's Cipher

The most known cipher goes as far back as to Julius Caesar in the antique Rome. Supposedly Caesar encrypted letters to Cicero and others by replacing the letters by other letters (substitution). It is said that his code was shifting the alphabet three letters to left, so that an „A" becomes a „D", a „B" becomes an „E" etc. Writing the two alphabets next to each other would lead to the translation table:

```
ABCDEFGHIJKLMNOPQRSTUVWXYZabcdefghijklmnopqrstuvwxyz

DEFGHIJKLMNOPQRSTUVWXYZabcdefghijklmnopqrstuvwxyzABC
```

Using this algorithm the first sentence of this chapter would read: `Wkh prvw nqrzq flskhu jrhv dv idu edfn dv wr Mxolxv Fdhvdu lq wkh dqwltxh Urph.` The structure of the text is still visible.

There is one basic problem with this algorithm: It does not change the statistical properties of the text. Each language has characteristic statistical properties like the probability of finding certain characters or character pairs etc. Figure 1 shows a histogram of the character distribution for three German and three English texts. Knowing these statistical properties it is straight forward to break the encryption. The most frequent character in both languages is the „E" (about 11.7% in English and 17.9% in German). Finding the most frequent character in the encrypted text, gives already the key (offset).

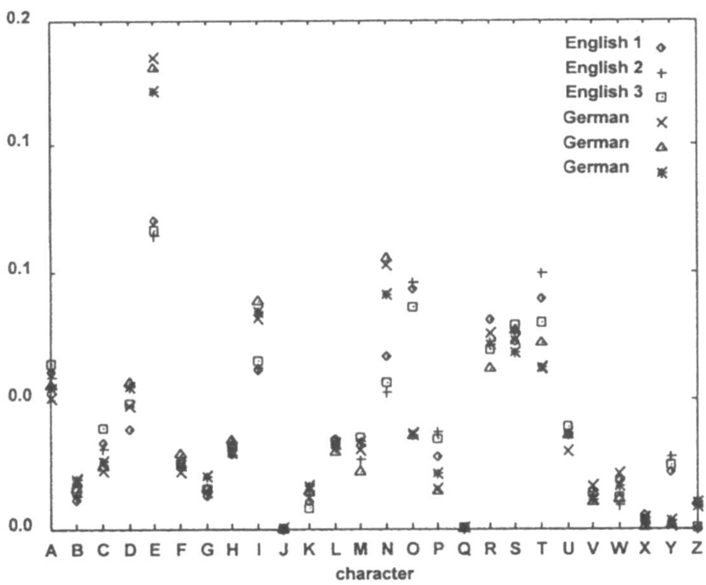

Figure 1: This figure shows a histogram of the distribution of the characters for three German and three English texts. Clear differences between the two languages and the characteristic patterns of the languages can be seen.

From this fact we can deduct that a good encryption algorithm has to destroy the statistical properties of the text completely.

2.2. Stream Ciphers

A simple way of encrypting a text so that all statistical properties are completely destroyed are the so called one-time pads. Each character is replaced by another character based on a random stream of characters. This basically means that for each character a different Caesar's Cipher is used (Vernam Cipher or one-time pad [2, 3]). And the random stream of characters determines which one. The problem of this is that the key (the random character stream) is as long as the text. The one-time pad, where the random characters were printed on a sheet of paper, is the old pre-computer version of this algorithm. The one-time pad is a perfect crypto-system, if the stream of characters is perfectly random [5]. Nowadays one would implement this using a pseudo random number generator.

In the computer age the alphabet has also only two characters, namely the bits „0" and „1". Implementing a stream cipher is done most effectively by XORing the text bitwise with a random stream of bits. A statistical physicist would do that by using a „good" random number generator in an encryption program like the one shown in Figure 2. Care must be taken, because a „good" sequence of random numbers is not necessarily a good sequence of random bits. Also pseudo random number generators never produce „real" random numbers. But it is an effective and simple way to write a small encryption program.

The statistical properties are destroyed as we can see in Figure 3. This figure is the result of encrypting one of the texts which was used to generate Figure 1. But the encryption is still breakable. Assuming that the size of the integer variables in the Listing from Figure 1 is 32 bits, the knowledge of 32 consecutive plain text bits and the corresponding 32 encrypted text bits is enough to break the code. A random number generator produces numbers with similar properties as random numbers, but nevertheless the sequence of numbers is deterministic and predictable. Knowing the random number generator algorithm and one random number is sufficient to know the next, the previous and all other random numbers.

Judging the Security of an algorithm we have to assume that the hacker knows the encryption algorithm. And one of the basic principles of cryptography is the so called Kerckhoffs Principle (A. Kerckhoffs von Nieuwenhof, 1835-1903): The Security of a cryptosystem must not depend on the secrecy of the algorithm. The Security of the cipher must reside entirely in the secret key. The DES algorithm is published since 1977 and practically not broken! The RSA algorithm is published since 1978 and not broken! If a software or Security system vendor has a good but secret algorithm do not trust him!

```
#include <stdio.h>
#include <stdlib.h>

#define KEY        0x12345678
#define BUFFERSIZE 1024

void error(char *text)
{
  printf("Error: %s\n",text);
  exit(1);
}

int main(int argc, char * argv[])
{
  FILE *infile, *outfile;
  size_t inread, outwrite ,i;
  char *buffer, *StartBuffer;

  if (argc != 3)
    error("Usage: CRYPT inputfile outputfile");
  printf("Input file:  %s\n",argv[1]);
  printf("Output file: %s\n",argv[2]);
  if ((infile=fopen(argv[1],"rb"))==NULL)
    error("Input File not found");
  if ((outfile=fopen(argv[2],"wb"))==NULL)
    error("Cannot open output file");
  if ((StartBuffer=(char*)malloc(BUFFERSIZE))==NULL)
    error("Cannot allocate buffer");
  srand(KEY);
  while(feof(infile)==0)
  {
    buffer=StartBuffer;
    inread=fread(StartBuffer,1,BUFFERSIZE,infile);
    for (i=0; i<inread; i++)
      *(buffer++) ^= (rand() & 0xff);
    outwrite=fwrite(StartBuffer,1,inread,outfile);
    if (outwrite != inread)
      error("Cannot write to output file");
  }
  free(StartBuffer);fclose(infile);fclose(outfile);
  return 0;
}
```

Figure 2: This listing gives a simple example of a ANSI C encryption and decryption program implementing a stream cipher. `rand` is a random number generator, who can be initialized using `srand`. The encryption key `KEY` is defined in the program. An improved version should read the key in. For Security reasons the key should not be supplied on the command line of the program. In multi user environments (e.g. UNIX) other users have access to the command line arguments.

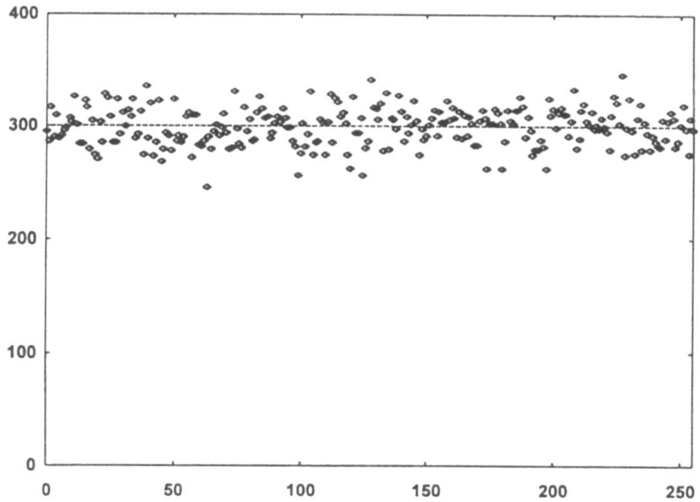

Figure 3: The result of encrypting an ASCII textfile using the program of Figure 2. The horizontal axis gives the ASCII code (0-255) of the characters and the vertical axis gives the character count. The horizontal line gives the theoretical result, if all character counts are the same.

2.3. DES

The DES algorithm is based on IBM's Lucifer algorithm. It was modified by the NSA and the actual design principle of core elements (strange tables called S-Boxes) is classified. This led to heavy criticism. But nevertheless is DES today an accepted standard. It was standardized by the National Bureau of Standards in 1977 [6] and recertified for five years in 1982, 1987 and 1992. Also it is known since about 22 years it is practically not broken.

The algorithm belongs to the class of Feistel ciphers. This type of ciphers splits a $2N$ Bit ($N=32$ for DES) word into a right half R_n and left half L_n of N bits each. The right half R_n becomes the left half of the next step $L_{n+1} = R_n$ and the left half is XORed with $f(R_n, K_n)$ before becoming the new right half R_{n+1}. Therefore the DES algorithm is given by:

$$
\begin{aligned}
L_1 &= IP(L_{in}, R_{in}) \\
R_1 &= IP(L_{in}, R_{in}) \\
L_{n+1} &= R_n \\
R_{n+1} &= L_n \text{ XOR } f(R_n, K_n), \quad n=1, 2, .., 15 \\
L_{out} &= IP^{-1}(L_{16} \text{ XOR } f(R_{16}, K_{16}), R_{16}) \\
R_{out} &= IP^{-1}(L_{16} \text{ XOR } f(R_{16}, K_{16}), R_{16})
\end{aligned} \qquad (1)
$$

IP is a initial permutation of the 64 input bits, which does not contribute to the Security of DES. It is obvious that several rounds are needed, because after one

Table 1: The table shows the speed of different commercially available DES encryption chips. [3]

Manufacturer	Chip	Speed
AMD	Am9518	10 MBit/sec
Safeware AG		10 MBit/sec
AMD	AmZ8068	14 MBit/sec
AT&T	T7000A	15 MBit/sec
Cryptech	Cry12C102	18 MBit/sec
CE Infosys	SuperCrypt 99C003	96 MBit/sec
VLSI Technology	VM007	192 MBit/sec
DEC		1 GBit/sec

round half of the bits are not encrypted at all. The DES algorithm uses 16 rounds. The 16 different 48 Bit keys K_n are generated by permutations and Bit shifts from the 56 Bit key, which is supplied by the user. It is not necessary that the function $f(R,K)$ has a inverse $f^{1}(R,K)$, because the structure of the Feistel network guaranties that, using the 16 keys in the opposite order, the algorithm can be inverted. Therefor the decryption algorithm is given by the same algorithm with the identical $f(R,K)$. The quality of such an algorithm depends on the proper choice of $f(R,K)$. There are numerous articles in cryptographic and computer publications [3, 4, 6, 7] which give the details of the function $f(R,K)$. They are not repeated here.

The DES algorithm was designed to be implemented in hardware and not in software. Therefor software implementations still do not perform good enough, so that they can be used. With the increase of the speed of typical processors used in PCs or workstations this will change. Typical performances of a few commercially available DES chips are given in Table 1.

2.4. RSA

All encryption algorithms we talked about up to now, use the same key for encryption and decryption. Therefore they are called symmetrical algorithms. The disadvantage of this is that everybody who has to read the text (and knows the key) can not only decrypt the text but also encrypt it. A solution to this problem was published 1976 by Diffie and Hellman in their work „*New Directions in Cryptography*“ [8]. They suggested to use a crypto system which uses for encryption (public key) and decryption (private key) to different keys. The keys are related, but the knowledge of one key does not allow to derive the other key. Today several realizations of such systems are known [2, 9].

In 1977 published Rivest, Shamir and Adleman [9] the most famous asymmetric crypto system: the RSA-algorithm. The RSA-algorithm is a direct application of Euler's theorem.

Let $F(n)$ be Euler's totient function, that means the number of positive integers $i<n$ such that $GCD(i,n)=1$. E.g. is $F(6)=2$, $F(10)=4$, and $F(15)=8$. If p is a prime, than

$$F(p)=p-1. \tag{2}$$

And if p and q are primes, than

$$F(pq)=(p-1)(q-1). \tag{3}$$

Euler's theorem now says that if $GCD(m,n)=1$ and $m<n$ then

$$m^{F(n)} \text{ MOD } n = 1. \tag{4}$$

An example for this is

$$13^{F(6)} \text{ MOD } 6 = 13^2 \text{ MOD } 6 = 169 \text{ MOD } 6 = 1 \tag{5}$$

Now let's see how the RSA-algorithm works. We chose two large primes p and q and multiply them

$$n = pq. \tag{6}$$

Then we search for two numbers e and d such that

$$ed \text{ MOD } F(n)=1. \tag{7}$$

The numbers e and d are the public and private key. The encryption of the message m is done by calculating

$$c=m^e \text{ MOD } n. \tag{8}$$

The decryption of the encrypted message c is done by

$$m=c^d \text{ MOD } n. \tag{9}$$

We see that the scheme works if

$$m^{ed} \text{ MOD } n = m \tag{10}$$

for all natural numbers $m<n$. The proof is straightforward using Euler's theorem [2,8]. Table 2 shows for $p=23$ and $q=37$ some choices for the keys e and d. Figure 4 shows a subroutine, which implements the RSA-algorithm for small numbers.

In order to break the algorithm, e.g. to calculate the private key d from the public key e and n, which are both known to everybody, one needs the number $F(n)$. To

Table 2: Some selected values for the keys *e* and *d* out of the 112 total possible pairs. The are constructed using *p=23* and *q=37*. This results in *n=851* and *F(n)=792*.

d	*e*	*d*	*e*	*d*	*e*	*d*	*e*
61	13	347	299	545	497	665	449
97	49	379	163	547	139	667	19
185	137	383	335	551	23	673	193
205	85	391	79	553	169	727	463
223	103	437	29	581	533	733	349
317	5	439	175	589	277	739	523
323	179	443	59	619	499	743	695
329	65	529	265	623	239	745	337
331	67	535	151	625	313	749	221
343	127	541	325	661	133	787	475

```
int RSA(int cm, int de, int
n)
{
    long t;
    int i;
    t=1;
    for (i=1;i<=de;i++)
      t=(t*cm) % n;
    return (int)t;
}
```

Figure 4: An implementation of the RSA algorithm in a ANSI C subroutine. For a real implementation long integer arithmetic must be used.

calculate *F(n)* one has to factor the number *n*. This is in principle possible, but it takes a very long time. In practice the number *n* is chosen such that it has 512 bits. This is large enough, so that it cannot be factored with today's computers in acceptable time (less then years CPU time).

In practice the RSA-algorithm is much to slow to be used as an crypto system for encrypting huge amounts of text. Therefor hybrid systems are in use. A text is encrypted with DES using a randomly generated key. This can be done fast. The DES-key is send along with the message being encrypted with a public RSA-key. Or to make sure that a text is not modified (e.g. a contract, bank transaction) a hash value [10] is calculated and the hash value is encrypted with the private key. A hash value is a kind of fingerprint, which uniquely identifies a electronic document. Now everybody can verify the integrity of the text by recalculating the hash value and comparing it with the decrypted hash value. This procedure is also called a digital signature.

RSA-keys are long (512 bits) and not chosen by a user. They cannot be remembered. Normally they are store on a floppy-disk or a smartcard, which is inserted in the proper reading device, when the key is needed. Such crypto systems are available commercially from different vendors. A good example is Zimmerman's package „Pretty Good Privacy" which is available for free [11].

3. Electronic Mail

The special problem of Electronic Mail (Email) is that the communication is not directly between sender and receiver. It is not known when sending a message, if the receiver's computer is active. Therefore always a mail server is involved in the delivery. The mail server can be a process running on the sender's computer. Especially with the a lot of different protocols (TCP/IP, X.400, Bitnet) there is nowadays a large tendency to set up one mail server per institution to take care of the protocols. For the sender of confidential mail it is not known in advance how the routing of its mail happens. The superuser of the mailserver can always read any mail he wants to read. It is pretty common in a university environment that a student is the superuser and a professor sends/receives confidential reports about students, sometimes even the superuser itself or close friends of him. This simply shows that there is no such thing as confidential Email. At least not with the standard protocols. There is action to extend the mail protocol (RFC 822) with „*Privacy Enhancement for Internet Electronic Mail*" (RFC 1421-1424).

4. Networks

Computer networks are considered to be a special Security risk. Let us look at the problems using the popular TCP/IP protocol as an example. This network uses not point to point connections, but a broadcast mechanism. A computer, who wants to transmit some data to another computer, sends them to the network. All computers on the same segment of the network read the data, to determine who is the addressee. Only the addressee uses the data, the other computer discard the data. All transmitted data including passwords are not send encrypted. With the proper software, which is available on the networks for free, the network card can be switched to promiscuous mode. This means that all the network traffic is logged by that computer. There have been incidents, where hacker used this feature to break sensitive passwords.

To make such a network more secure the network cable must be put in areas (cable channels), where it is not accessible by unauthorized people. Care must be taken if remote computers have to be accessed. One has to realize that a remote login, e.g. via a telnet session, requires the transfer of the unencrypted password through the network.

4.1. Passwords

The most simple way of storing passwords is the storage of the plain unencrypted password. Surprisingly this is still done in current products. The latest known case is the storage of the passwords in file CCMAIL.CFG by the program cc:Mail for Windows from Lotus [12]. A slight improvement over this is the storage of an encrypted password, but it is decrypted for comparison with the user supplied password. With some knowledge of the software it is possible to get access to the decrypted password while it is in memory. Much better is the solution, which is implemented in UNIX. The password is stored encrypted using a one-way algorithm and never decrypted. The user supplied password is also encrypted and the two encrypted passwords are compared. This feature clearly shows a way for improvement. Any remote login process should require that the password is already encrypted, before the password is send.

Care must be taken that no unauthorized user can access the encrypted passwords in the password file. It is easy to crack the passwords by using a dictionary of common passwords, which are encrypted and then compared with the passwords in the password file. This attack has been successful in the case of weak passwords (e.g. first name of spouse, friend's name, „Obelix“, userid backwards, etc.). To force the users not to use weak passwords, it is a good idea to run a password cracking program permanently in the background.

Some common sense is to be used in choosing „good“ passwords. Passwords should be at least six characters and should contain at least two non alphabetic characters. If passwords are chosen to complicated users tend to write them down, which is a major Security hole.

4.2. Challenge Response Protocols

A major improvement is getting away from a simple static password authentication. Using different data for authentication every time the information obtained from eavesdropping the network is useless. MIT's Kerberos [13] is a good system which implements this concept. The authentication is handled by a challenge-response protocol. This guaranties that at no time a password is transferred in the network.

The computer, who initiates the authentication, send a random number *RAND* to the user. The user encrypts this random number using his password k and the encryption algorithm f.

$$RES=f_k(RAND) \tag{11}$$

The encrypted random number *RES* is send back to the computer. The computer knows the user password k' so that it can encrypt the random number, too.

$$RES'=f_{k'}(RAND) \tag{12}$$

The computer compares *RES* with *RES'* and if they are equal it knows that $k=k'$. Using this protocol the user password is never transferred in clear over the network.

The calculation of *RES* by the user is complicated. Therefore it is not easy to use this for a login from a terminal. Modern systems use personalized hardware devices for this. One possibility is the use of smartcard. Smartcards are creditcard sized plastic cards, which contain a one chip microcomputer. The smartcard is inserted into a reader and performs the authentication for the user. In order to use the smartcard the user has to give a password to the smartcard. If the reader is built into a keyboard this password never leaves the keyboard. The whole procedure is for the user very similar to using a banking card with a banking machine.

Another solution is the use of a handhold device like a pocket calculator. The user reads a number from the screen (challenge), types it together with his password into the little device. After this he can read a number on display (response) which he types into the computer. This is more complicated for the user and the smartcard solution is preferable.

5. Stand-alone PC

Personal computers are the most common computers, which are used everywhere. Physicians, lawyers, counselors, scientists, and lots of other people store most sensitive data on of the shelf personal computers. If such a computer breaks down, and is brought to a repair shop, this data are exposed to an unpredictable number of people. Even the BIOS password of a PC can be easily circumvented by putting the harddisk into another computer and it is read there. The only protection of data is encryption of the data. Several vendors offer software and hardware systems to encrypt files on the harddisk of a PC. Even some applications (typically spreadsheets and word-processing programs) over built in encryption. Earlier this year it was reported, how easy it is, to break the software encryption of main stream spreadsheets [14].

5.1. Software Encryption

Software encryption has the problem of performance. Especially on-line encryption of communication or data storage devices is very demanding on the encryption speed. For this reason all vendors of software solutions tend to use speed optimized algorithms, which normally means a variation of a XOR encryption algorithm. This algorithms are so weak, that they can be broken automatically [14, 15].

There exist now encryption algorithms, which are as good as DES but much faster. A good algorithm is Lai and Massey's IDEA algorithm[4, 16]. A pretty new one is Schneier's „Blowfish“ algorithm [17], which is simple to implement on 32 bit microprocessors. The package „Pretty Good Privacy“ shows that a serious application can be implemented in software. It uses RSA for key encryption and key exchange and IDEA for text encryption. But the price to pay is performance. The solution to this is a hardware implementation.

5.2. Hardware Encryption

Quite a number of encryption chips are available today (see Table 2). The Clipper Chip project of the United States NSA wants to implement a crypto chip into every telephone, computer modem and fax machine. The fact that the chip contains a „back door" to allow government agencies access to encrypted communication created quite some criticism and discussion about the privacy of communication in the U.S.A. [18].

Using this hardware technology a Security system with almost no downgrade in performance of a computer can be installed. Today of the shelf Security hardware systems are available. There are devices, which encrypt communication lines, or devices, which encrypt entire harddisks in PCs.

With the open operating systems of PCs a reliable user authentication and access control is only possible with encryption of the harddisks. While the Security system is installed it controls access, as soon as it is deinstalled the data on the disks are not accessible. A typical system consist of a hardware board with a DES chip and a smartcard reader for user authentication. The typical cost is between 500 US$ and 1000 US$.

Acknowledgment. I want to thank M. Schrempf for valuable suggestions and for carefully reading the manuscript.

References

1. W. Madsen, *Handbook of Personal Data Protection* (Stockton Press, New York, 1992).
2. A. Beutelspacher, *Kryptologie* (Vieweg, Braunschweig, 1993).
3. W. Fumy and H.P. Rieß. *Kryptographie* (Oldenburg, München, 1994).
4. B. Schneier, *Applied Cryptography: Protocols, Algorithms and Source Code in C* (John Wiley & Sons, New York 1994).
5. C.E. Shannon, Bell. Sys. Tech. J. **30**, 657 (1949).
6. National Bureau of Standards: *Data Encryption Standard* FIPS Publication 46, 1977.
7. W.H. Press, B.P. Flannery, S.A. Teukolsky, and W.T. Vetterling, *Numerical Recipes* (Cambridge University Press, Cambridge 1986).
8. W. Diffie and M. Hellman, Trans. IEEE Inform. Theory, **IT-22**, 644 (1976).
9. R. Rivest, A. Shamir and L. Adleman, Comm. ACM **21**, 120 (1978).
10. B. Schneier, Dr. Dobb's Journal, September 1991, page 148; W. Stallings, Dr. Dobb's Journal, April 1994, page 32.

11. P. Zimmermann, *Pretty Good Privacy*. Available via anonymous ftp on the internet for DOS, UNIX and Macintosh.
12. H.J. Highland, Computers & Security **13**, 101 (1994).
13. J.G. Steiner, C. Neuman and J.I. Schiller, USENIX Workshop Proceedings, UNIX Security Workshop, Portland OR, 1988.
14. PC Welt, February 1994, page 92 and March 1994, page 153.
15. c't - magazin für computer technik, June 1994, page 184.
16. X. Lai and J.L. Massey, *Advances in Cryptology - Eurocrypt '90*, Springer Lecture Notes in Computer Science **473**, 389 (1991); X. Lai, J.L. Massey and S. Murphy, *Advances in Cryptology - Eurocrypt '91*, Springer Lecture Notes in Computer Science **547**, 17 (1991)
17. B. Schneier, Dr. Dobb's Journal, April 1994, page 38.
18. Time Magazine, March 14, 1994, page 38.

Part IV

Contributed Papers

Monte Carlo with Absorbing Markov Chains: A Faster Monte-Carlo Algorithm for Dynamical Studies

M.A. Novotny

Supercomputer Computations Research Institute, Florida State University, Tallahassee, FL 32306-4052, USA

Abstract. A new Monte Carlo algorithm is introduced which combines Monte Carlo with absorbing Markov chains (MCAMC). With a few additional assumptions, the simplest form of the MCAMC algorithm corresponds to the n-fold way algorithm. The MCAMC algorithm is applied to the study of the escape from the metastable state of the square-lattice Ising ferromagnet. Results agree with both continuum and discrete droplet theory.

1. Introduction

Although there are many Monte Carlo algorithms to efficiently calculate static quantities of model statistical mechanical systems [1,2], there is at present only one Monte Carlo algorithm to efficiently calculate dynamical quantities. This is the n-fold way algorithm introduced almost two decades ago [3]. In this paper, a new Monte Carlo algorithm, the MCAMC algorithm, that combines absorbing Markov chains and Monte Carlo simulations will be introduced. The efficiency of the MCAMC algorithm will be demonstrated by applying it to the escape from the metastable state of the nearest-neighbor Ising model [4–6].

2. The MCAMC Algorithm

The Markov matrix for an Absorbing Markov Chain (AMC) with q absorbing states and s transient states is given by [7]

$$\mathbf{M} = \begin{pmatrix} \mathbf{I}_{q\times q} & \mathbf{0}_{q\times s} \\ \mathbf{R}_{s\times q} & \mathbf{T}_{s\times s} \end{pmatrix} \tag{1}$$

where the subscripts indicate the submatrix size. The matrix $\mathbf{I}$ is the identity matrix, $\mathbf{0}$ is the zero matrix, $\mathbf{R}$ is the absorbing submatrix with elements equal to the transition probabilities from the transient to the absorbing states, and $\mathbf{T}$ is the transient submatrix with elements equal to the transition probabilities between transient states. Starting from an initial vector $\vec{v}_I$, the probability of still being in one of the s transient states after m time steps is given by $\vec{v}_I^T \mathbf{T}^m \vec{e}$, where $\vec{e}$ is the vector with all elements equal to unity. Using a random number r which is uniformly distributed between zero and one, one can obtain a time m for exiting to one of the q absorbing states by the solution of the equation

Springer Proceedings in Physics, Vol. 78
Computer Simulation Studies in Condensed-Matter Physics VII
Editors: D.P. Landau · K.K. Mon · H.-B. Schüttler

$$\vec{v}_I^T \mathbf{T}^m \vec{e} < r \leq \vec{v}_I^T \mathbf{T}^{m-1} \vec{e}. \tag{2}$$

Note that m is independent of the particular state to which the system exits. The elements of the vector of unnormalized probabilities

$$\vec{v}_I^T \left(\mathbf{I} + \sum_{k=1}^{m-1} \mathbf{T}^k \right)_{s \times s} \mathbf{R}_{s \times q} \tag{3}$$

give the probability of exiting to each of the q states. Another random number r can be used to pick which state the system ends up in. Eq. (2) and (3) are the only equations necessary to utilize AMC in a Monte Carlo simulation.

If s=1 the Markov matrix corresponds to absorbtion taking place whenever the system exits from the current state. If the system has N spins, the dynamics desired is to randomly choose a spin to flip and then decide whether or not to flip it. For discrete spin systems, the number of possible energy classes a spin can have is an integer n. Let n_i be the number of spins in the current spin configuration that are in class i, and $p(i)$ be the probability of flipping a spin in class i. Then the absorbing submatrix is given by $\mathbf{R}_{1\times n} = \big(n_1 p(1) n_2 p(2) \cdots n_n p(n)\big)/N$. Define $Q_j = \sum_{i=1}^{j} n_i p(i)$, and Q_0=0. Since $\mathbf{M}$ is a Markov matrix, its row sums are equal to one. Hence the transient submatrix is $\mathbf{T}_{1\times 1} = 1 - Q_n/N$. A spin in class j is chosen to flip if a random number r satisfies $Q_{j-1} < rQ_n < Q_j$. If class j has more than one spin, a random integer $\in [1, n_j]$ is chosen and this spin flips. This corresponds to the general case given by Eq. (3). The time increment m taken to flip this spin is found using a random number r and $m > \ln(r)/\ln(1-Q_n/N) > m-1$. The general case of this relationship is given by Eq. (2). For small Q_n/N, the expansion $1/\ln(1-Q_n/N) \approx -N/Q_n$ can be employed, and m can be treated as a continuous time step. Then the time increment for the chosen spin to flip is given by $m = -N\ln(r)/Q_n$. This corresponds to the traditional n-fold way algorithm [3].

3. Metastability in the Ising Model

The nearest-neighbor (nn) square-lattice Ising model has the Hamiltonian $\mathcal{H} = -J\sum_{\langle i,j \rangle} \sigma_i \sigma_j - H \sum_i \sigma_i$ where $\sigma = \pm 1$ and the first sum is over nn spins. With periodic boundary conditions, the n=10 classes of spins are ordered as 1–5 (6–10) according to if the spin is +1 (−1) with 4–0 (4–0) nn +1 spins [3].

Although the MCAMC algorithm is applicable to other dynamics, in this study only the Metropolis dynamic is used. A spin to be tested is randomly chosen, and is always flipped if the energy E_{new} of the new configuration is lower. Otherwise it is flipped with the Boltzmann probability $\exp\big((E_{\text{old}} - E_{\text{new}})/T\big)$.

To study metastability, start with the initial configuration to have all spins +1, and the applied magnetic field H to be negative. As shown in Fig. 1, at low temperatures the n-fold way produces a substantial decrease in computer time compared to the normal Monte Carlo algorithm. However, the MCAMC algorithm from Eq. (2) and (3) allows for another substantial decrease in computer time. For s=2 MCAMC the AMC corresponds to the system being in

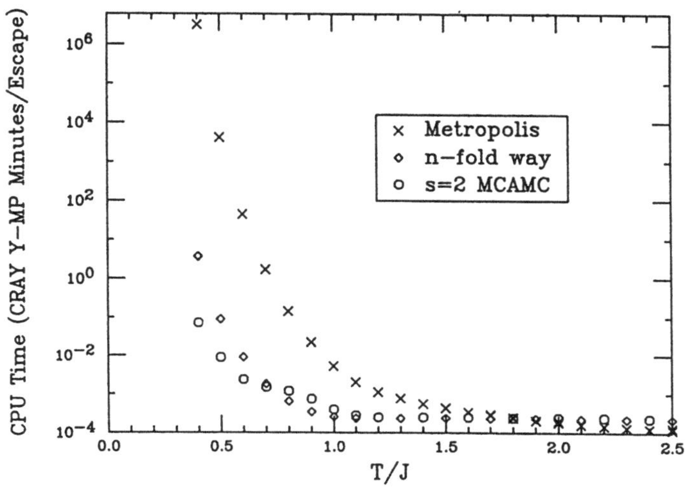

Fig. 1 The average CPU time per escape from the metastable state is shown as a function of temperature for a 24×24 lattice with $J=1$ and $H=3J/4$. The timings are for a single processor Y-MP. The n-fold algorithm is equivalent to $s=1$ MCAMC. For small T the timings for the × points are estimates obtained using the average lifetime τ obtained with the MCAMC algorithm.

the initial state and exiting into a state with two overturned spins. (The n-fold way algorithm has $s=1$ with the initial configuration exiting to a state with one overturned spin. However, at low temperatures this overturned spin often is chosen to flip and the system reverts to the initial state.) The nontrivial submatrices in Eq. (1) are

$$\mathbf{R} = \begin{pmatrix} 0 & 0 \\ \frac{4p(2)}{N} & \frac{(N-5)p(1)}{N} \end{pmatrix} \qquad \mathbf{T} = \begin{pmatrix} 1-p(1) & p(1) \\ \frac{p(6)}{N} & 1-\frac{S}{N} \end{pmatrix} \tag{4}$$

with $S=p(6)+4p(2)+(N-5)p(1)$. The first transient state is the initial state, and the second is the state with one overturned spin. The first absorbing state corresponds to two nn overturned spins, and the second to two overturned spins that are not nn spins. In the MCAMC algorithm, if the configuration has two or more overturned spins, the normal n-fold way algorithm is used. If the system returns to the initial state, the AMC part of the algorithm is again used. Fig. 1 shows that at low temperature the $s=2$ MCAMC algorithm is much faster than the n-fold way algorithm. For a larger s, the MCAMC algorithm will yield an even faster algorithm at low temperatures and fields.

From continuum droplet theory, for weak H the lifetime τ of the metastable state is given by [6,8–10]

$$\tau = A(T)|H|^{-b-c}\exp\left(\Xi(T)/|H|\right) \tag{5}$$

where $\Xi(T)$ is related to the surface tension and the equilibrium spontaneous

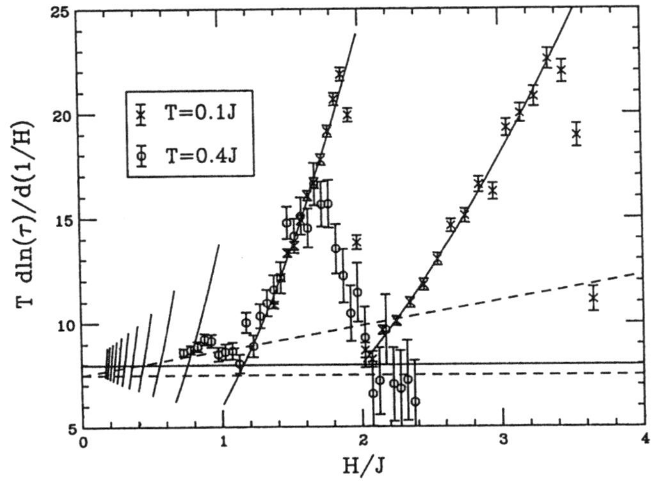

Fig. 2 The derivative of $T\ln(\tau)$ with respect to $1/H$ is shown as a function of H for a 24×24 lattice. The horizontal lines are the exact zero-field values of $T\Xi(T)$ [12] for T=0.1J (solid) and T=0.4J (dashed). The other dashed line is for T=0.4J and has the form from Eq. (5) using the exact zero-field value of Ξ and the theoretical value b+c=3. The solid curves are the results predicted from Eq. (6) from low-temperature discrete droplet theory. The values shown are averages over 10^3 escapes from the metastable state, and the numerical derivatives of $\ln(\tau)$ were taken using a three-point formula with ΔH=0.25.

magnetization. A field-theoretical calculation gives the prefactor exponent associated with 'Goldstone-mode' fluctuations to be b=1 [9]. The prefactor exponent c gives the H dependence of the 'kinetic prefactor' [10].

At low temperatures and high fields the discreteness of the nucleating droplets becomes important. As $T\to 0$, the lifetime is given by [11]

$$k_B T\ln(\tau) = \Gamma(H,J) = 8J\ell_c - 2H(\ell_c^2 - \ell_c + 1) \tag{6}$$

where ℓ_c=[2J/H]+1 and the square brackets denote the integer part. This result is restricted to $2J/H$ not an integer and to H<4J.

Fig. 2 shows that at low temperatures and intermediate fields the discrete droplet results of Eq. (6) adequately predict the behavior. At stronger fields, the single-droplet nucleation picture which leads to Eq. (6) no longer holds. The 'dynamical spinodal', H_{DSP}, separates the single and multidroplet decay regions [5,6]. H_{DSP} depends on the system size as well as on T. In Fig. 2 only fields weaker than the estimate for H_{DSP}, which was obtained from the location where the standard deviation of the escape time τ is $\tau/2$, have been plotted. At weak H for T=0.4J, Fig. 2 shows that the continuum droplet prediction of Eq. (5) is approached. The weakest fields shown in Fig. 2 required about 10 CRAY Y-MP CPU hours per H value.

4. Conclusions

A new algorithm which incorporates Monte Carlo and absorbing Markov chains (MCAMC) has been introduced. The MCAMC algorithm can be many orders of magnitude faster in computer time than traditional Monte Carlo algorithms.

Results for escape from the metastable state for the square-lattice nn Ising ferromagnet show that for high enough temperature and weak enough fields predictions from continuous droplet theory hold, whereas for low temperatures and high fields predictions from low-temperature discrete droplet theory hold.

The MCAMC algorithm should also be applicable to other situations where the interesting physics is in the slow relaxation of the system, but where there are some very fast modes in the system. Examples include late-time coarsening phenomena and dynamics of spin-glass models. Currently, the MCAMC algorithm is being utilized to test the prediction [13] that at low T the metastable state of the anisotropic square-lattice nn Ising model decays through nucleation of square droplets, rather than equilibrium Wulff droplets.

Acknowledgements.

Supported in part by the Florida State University Supercomputer Computations Research Institute, which is partially funded through Contract #DE-FC05-85ER25000 by the U.S. Department of Energy, and by Florida State University through time granted on its CRAY Y-MP supercomputer. Discussions with C. C. A. Günther, J. Y. Lee, and P. A. Rikvold are acknowledged.

References

[1] R. H. Swendsen and J. S. Wang, Phys. Rev. Lett. **58**, 86 (1987).
[2] B. A. Berg, in these proceedings.
[3] A. B. Bortz, M. H. Kalos, and J. L. Lebowitz, J. Comput. Phys. **17**, 10 (1975).
[4] For a review see K. Binder, in *Phase Transitions and Critical Phenomena*, eds. C. Domb and M. Green (Academic, London, 1976), Vol. 5B.
[5] H. Tomita and S. Miyashita, Phys. Rev. B **46**, 8886 (1992).
[6] P. A. Rikvold, H. Tomita, S. Miyashita, and S. W. Sides, preprint FSU-SCRI-94-11.
[7] M. Iosifescu, *Finite Markov Processes and Their Application*, p. 99 (John Wiley & Sons, New York, 1980).
[8] M. E. Fisher, Physics **3**, 255 (1967).
[9] N. J. Günther, D. A. Nicole, and D. J. Wallace, J. Phys. A: Math. Gen. **13**, 1755 (1980).
[10] J. S. Langer, Ann. Phys. **54**, 258 (1969).
[11] E. J. Neves and R. H. Schonmann, Commun. Math. Phys. **137**, 209 (1991).
[12] R. K. P. Zia and J. E. Avron, Phys. Rev. B **25**, 2042 (1982).
[13] R. Kotecky and E. Olivieri, J. Stat. Phys. **70**, 1121 (1993).

Non-Arrhenius Conductivity in a Driven System of Interacting Lattice Gas

R.B. Pandey

Department of Physics and Astronomy, Program in Scientific Computing, University of Southern Mississippi, Hattiesburg, MS 39406-5046, USA

Abstract: A Monte Carlo simulation is used to study the conductivity of an interacting lattice gas in presence of a linear gradient field. The conductivity is evaluated in the steady state although the system is not in thermal equilibrium. Effects of the concentration of the carriers and the temperature are analyzed. The conductivity shows a nonmonotonic dependence on the concentration and a non-Arrhenius dependence over a wide range of temperatures.

Processing of various additives into fluid, polymers, ion separation and filtering through membranes in a variety of aqueous solutions are some of the common examples which require a good understanding of the transport properties of the constituent particles i.e. ions, ionomers and neutral particles [1-4]. Experimental methods [1-3] such as NMR, light scattering, neutron scattering, and dielectric measurement techniques are frequently employed to study conductivity, diffusion, polarization, and dielectric response. These transport properties are very often caused by biased fields which may be set up by an external field, the distribution of constituent ions (charge density gradient) or due to combination of both. One of the major problems in studying the transport properties in such complex systems is to find suitable methods to handle the interactions of various range. Although, the study of the transport properties in lattice gas has attracted a considerable interest for a long time [5,6], such investigations for the interacting lattice gas systems have begun to intensify only recently, particularly in a driven field [7,8]. We present here a computer simulation study of the transport properties of an interacting lattice gas in presence of a linear gradient field.

We consider a discrete lattice. A fraction p of the sites is randomly occupied by particles of type A, each with unit charge (interaction) density ($\rho_A = 1$); a lattice site cannot be occupied by more than one particle (the hard-core interaction). Each empty site (called a B-type particle) is assigned a negative density ($\rho_B = -p/(1-p)$) such that the total charge of the whole lattice is zero. Each lattice site is thus assigned an interaction density. We consider a limited range of Coulomb-like interactions among these particles with a linear gradient biased field, i.e. the interaction energy is given by,

$$H = \sum_{ij} \rho_i \rho_j / r_{ij} + \sum_{ix} B_0(1-ix/L) , \qquad (1)$$

where ρ_i is the charge density of site i, r_{ij} is the distance between site i and j, ix is the x-coordinate of site i and L is the linear size along the x-direction; the biased field strength B_0 is set unity. We use Metropolis algorithm to move a particle to one of its neighboring empty sites chosen randomly (i.e. the exchange of particles A and B) with a periodic boundary condition.

Transport quantities such as the power-law dependence of the root mean square (rms) displacement of mobile particle and that of their center of mass, domain formations and related transition, temporal evolution of energy, and flux of particles as a function of time are studied in two and three dimensions. Effects of temperature and concentration on these quantities are explored in detail. Many unusual transport properties such as non-diffusive

Springer Proceedings in Physics, Vol. 78
Computer Simulation Studies in Condensed-Matter Physics VII
Editors: D.P. Landau · K.K. Mon · H.-B. Schüttler

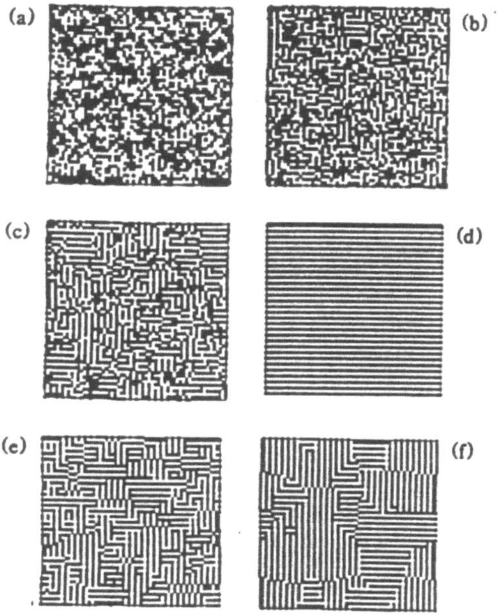

Figure 1. Snapshot of the configurations in steady state (at 10^5 MCS) with the next nearest neighbor interaction at temperature T = 10.00 (a), 1.00 (b), 0.50 (c), 0.30 (d), 0.10 (e), and 0.05 (f) on a 60X60 sample. A particle is shown by filled circle, and empty site (hole) by a open space except in a perfectly mixed microstate (filled square) in which either a particle is surrounded by all neighboring holes or a hole surrounded by all neighboring particles. (ref.7).

power-law for the rms displacements, growth and decay of domain of A-B particles, and conductivity are found [7]. In two dimensions it is relatively easier to visualize the cluster formation and domain growth. Since our model is an analog of antiferromagnetic model with the Kawasaki dynamics (in some concentration limit i.e. p = 0.5 with a nearest neighbor interaction), the onset of infinite cluster, a characteristic of an antiferromagnetic ordering is not unexpected; we have seen such an ordering with the nearest neighbor and fourth neighbor interactions [7]. However, with the next nearest neighbor interaction, we have observed an interesting pattern in the growth and decay of domains as a function of temperature, as shown in figure 1.

From the plots of energy versus time, we find that the energy decays with time constantly as the system is driven out of equilibrium by the gradient field. However, the flux of particles at the opposite sides along the x-direction, shows a steady state behavior (see figure 2). These characteristics lead to a non-equilibrium steady state (NESS) system; various NESS systems are studied in recent years which seems to exhibit unusual features [7,8].

It is rather easy to estimate the conductivity from the flux Q along the gradient field in our model in the steady state. The current density j is given by,

$$j = (1/A)\, dQ/dt \;, \qquad (2)$$

where j is the current density along the x-direction and A is the cross-sectional area. Using, $j = \sigma E_g$, where σ is the conductivity and E_g is the gradient field (see eq. 1), we can evaluate σ.

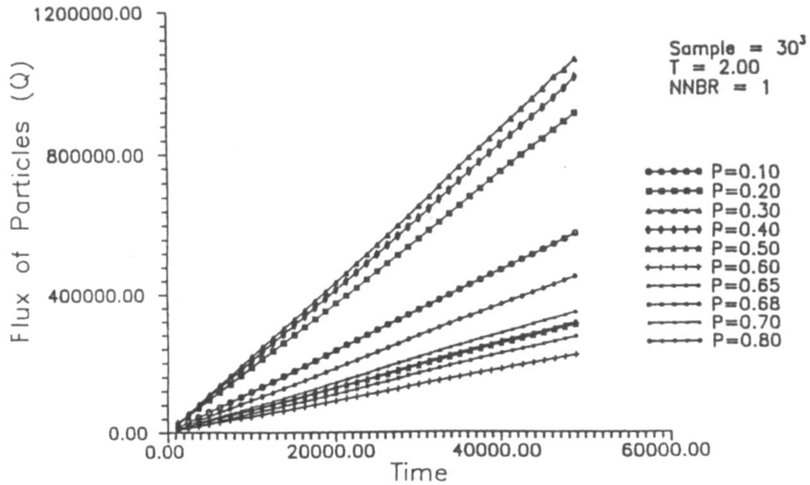

Figure 2. Particles' flux versus time for the nearest neighbor interaction (NNBR = 1) for various concentrations (p = 0.10 - 0.80) at T = 2.00.

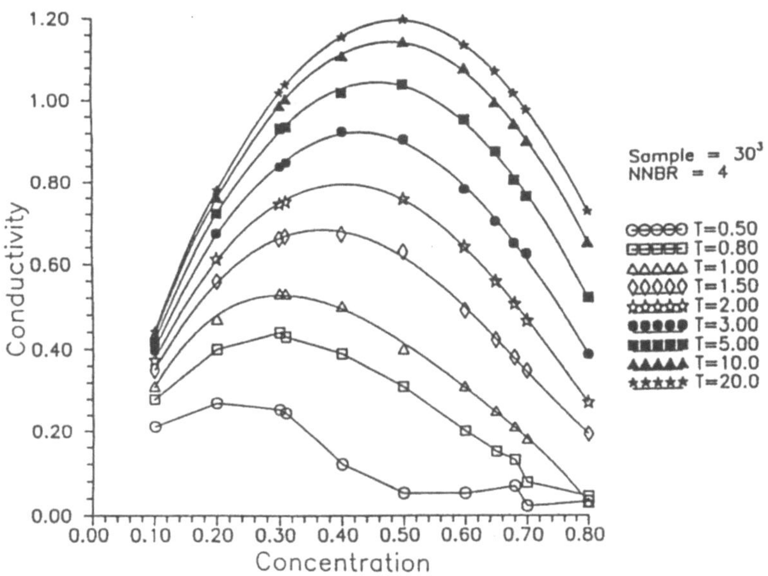

Figure 3. Conductivity versus concentration at different temperature for the range of interaction up to fourth neighbors (NNBR = 4).

We have studied in detail the variation of the conductivity with the concentration p of the particles (carriers) and the temperature. Figure 3 shows a typical plot of σ versus p. We see that the conductivity depends nonmonotonically on the concentration. At the high temperatures (T = 1.00 - 20.00), the conductivity shows a maximum which seems to shift toward lower concentrations on reducing the temperature. On further decreasing the

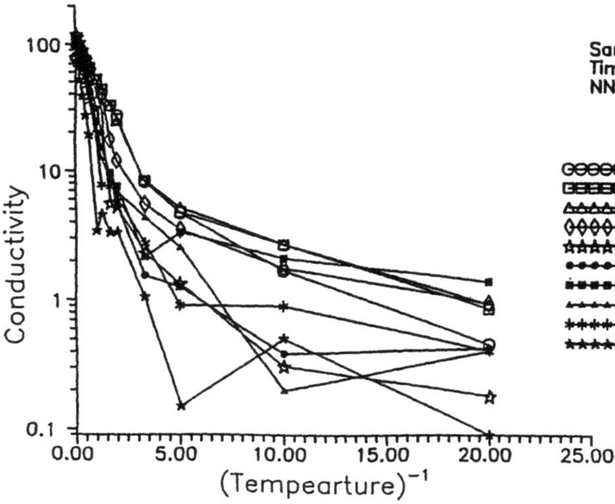

Figure 4. Conductivity versus 1/T on a semilog scale for NNBR = 4 at various concentrations.

temperature below, T = 1.00, a minima begins to emerge which is visible at T = 0.50 at p = 0.50. Thus, the conductivity shows two maxima separated by the minima at T below a characteristic value T^* (~ 1.00) above which this minima vanishes.

The Arrhenius dependence of conductivity on the temperature is given by $\sigma \sim \exp(-E_a/T)$ where E_a is the activation energy i.e. the conductivity versus 1/T on a semilog scale should show a linear dependence. From figure 4, we see that such dependence holds only at high temperature. The conductivity deviates strongly from the Arrhenius dependence at low temperatures particularly at higher concentrations. A detail study will be published elsewhere.

Acknowledgements: Author would like to acknowledge a partial support from the NSF-EPSCoR grant. Computer time was provided by the Mississippi Center for Supercomputing Research.

References:

[1] K.A. Mauritz and R.M. Fu, Macromolecules 21, 1324 (1988); H. Yoon, H. Kim, and H. Yu, Macromolecules 22, 848 (1989); M. Hara, J. Yu, and A.H. Lee, ibid. 22, 754 (1989).

[2] C.W. Lantman, W.J. Macknight, D.G. Peiffer, S.K. Sinha, and R.D. Lundberg, Macromolecules 20, 1096 (1987).

[3] T. Sakai, H. Takenaka, N. Wakabayashi, Y. Kawami, and E. Torikai, J. Electrochem. Soc. 132, 1328 (1985).

[4] H. Bottger and V.V. Bryksins, Phys. Status Solidi B 113, 9 (1982).

[5] N. Wax (ed.), "Selected Papers on Noise and Stochastic Processes" (Dover, New York, 1954); E. W. Montroll and J.L. Lebowitz (eds.) "Fluctuation Phenomena" (North-Holland, Amsterdam, 1979).

[6] J.W. Haus and K.W. Kehr, Phys. Rep. 150, 263 (1987).

[7] Y. He and R.B. Pandey, Phys. rev. Lett. 71, 565 91993); also in "Computer Simulation Studies in Condensed Matter Physics VI" eds. D.P. Landau, K.K. Mon, and H.-B. Schüttler (Springer-Verlag, 1993).

[8] B. Schmittmann, K. Hwang, and R.K.P. Zia, Europhys. Lett. 19, 19 (1992); B. Schmittman and R.K.P. Zia, Phys. Rev. lett. 66, 357 (1991); S. Katz, J.L. Lebowitz, and H. Spohn, Phys. Rev. B 28, 1655 (1983).

Non-Equilibrium Steady States of Ising Systems

K.E. Bassler

Center for Stochastic Processes in Science and Engineering and
Physics Department, Virginia Polytechnic Institute and State University,
Blacksburg, VA 24061-0435, USA

Abstract. We examine the results of some recent simulations concerning the critical behavior in the nonequilibrium steady states (NESS) of various Ising systems. Unlike equilibrium systems, where the static properties are independent of the dynamics, the behavior of nonequilibrium systems depends crucially on the dynamics. Various NESS Ising models, having both conserved and non-conserved dynamics, are considered. Their phase diagrams and the universality classes of their critical properties are discussed.

1. Introduction

Statistical mechanics has proved to be a powerful tool useful for understanding the collective behavior of many-body systems. Much of our understanding is based on equilibrium Statistical mechanics which has been developed over the last century. However, most natural phenomena is inherently nonequilibrium, and our understanding of nonequilibrium statistical mechanics is tenuous, for a complete understanding of nonequilibrium statistical mechanics is very difficult to achieve. We do have some understanding of the behavior of systems near equilibrium, where the nonequilibrium behavior is only transient and can be considered as a perturbation from the equilibrium steady state, but, far from equilibrium, where, in general, the steady state can not be defined in terms of a Boltzmann distribution, much less is known.

In an effort to study far from equilibrium systems, there has recently been considerable interest in the statistical mechanics of simple models, such as interacting lattice gases, which have been driven into nonequilibrium steady states (NESS) by external forces. In those systems, a variety of interesting behavior has been discovered including singular correlations at all temperatures, and critical points with novel universal behavior [1].

In equilibrium systems, the only essential features that affect the "static" universal critical behavior of the system are the symmetry and the dimensionality of the system [2]. On the other hand, in nonequilibrium systems the dynamics itself is also important, and so the natural question is: What are the essential features of the dynamics that affect the universal critical behavior of nonequilibrium systems?

In order to attempt to address this question it is useful to study the simplest nontrivial example of a system with a phase transition, which is, of course, the Ising model. The Ising model [3] consists of classical spins $S_i = \pm 1$ confined to the i sites of a regular lattice, and which has an energy

$$H[C] = -J \sum_{\langle ij \rangle} S_i S_j \quad , \tag{1}$$

where the sum runs over all nearest neighbor pairs of lattice sites.

In this paper I discuss some recent Monte Carlo simulation results concerning the critical behavior and the topology of phase diagrams in NESS Ising systems. A variety of such systems have been studied. I will not attempt to describe all of them, but instead will concentrate on a few that give some indication what the essential features of the dynamics are that influence the universal critical behavior.

2. Phenomenology

The dynamics used in the studies we discuss below consist of either spin-flip, Glauber type dynamics [4], which does not conserve the total magnetization of the system, or spin-exchange, Kawasaki type dynamics [5], which does conserve the total magnetization of the system. The transition to ferromagnetic order is very different for the two types of dynamics. In a spin-flip system, an ordered state has non-vanishing magnetization, but in a spin-exchange system since the total magnetization is conserved, the transition to ferromagnetic order appears as phase separation; the system separates into a + spin rich and + spin poor domain. In equilibrium, the static critical properties are same for the different dynamics. However, as we will see, away from equilibrium differences in the dynamics can produce very different critical behavior. All of the models discussed below were simulated on square lattices with periodic boundary conditions, and all spin-exchange systems were restricted to zero total magnetization. Furthermore, we also restrict our consideration to ferromagnetic coupling $J > 0$, and without loss of generality set $J = 1$.

Among the NESS Ising systems having spin-flip dynamics that have been studied using Monte Carlo methods are a collection of models known as two temperature models. Two temperature models are coupled to two different heat baths, which in general are at different temperatures. There is an energy current that flows from the higher temperature heat bath to the lower temperature heat bath, and this energy flux "drives" the system away from equilibrium. The steady states of two temperature models are thus non-equilibrium, and generally can not be given in terms of a Boltzmann distribution.

For example, Blöte, et. al. have studied a "checkerboard" spin-flip two-temperature model [6]. In that model, the sites of the square lattice are divided into "white" and "black" sites in a checkerboard pattern. The spins on a white (black) sight are coupled to the "white" ("black") heat bath and a spin flip occurs at the temperature of that heat bath, e.g., a spin at site i will flip with the Metropolis probability $P = \min\{1, \exp(\beta_i \Delta H)\}$, where β_i is the inverse temperature of the heat bath coupled to site i. A similar model, which

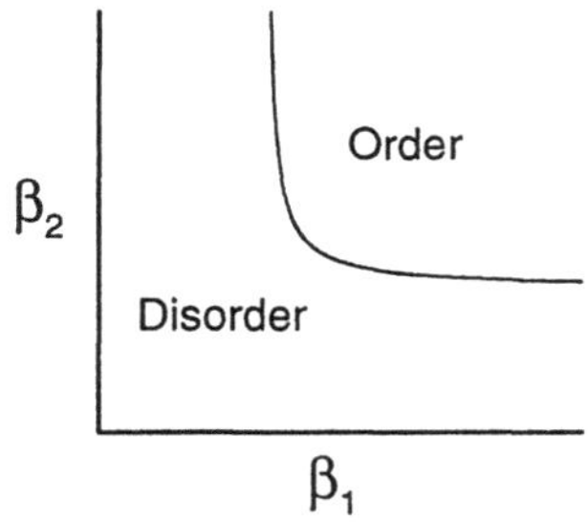

Fig. 1: Phase diagram for spin-flip two temperature models. Dimensionless inverse temperatures are denoted by $\beta_i = J/T_i$. The line separating the disordered and ordered phases is a line of continuous transitions.

has also been studied by Blöte, et. al, is the "strip" spin-flip two temperature model [6], where again the sites of the square lattice are divided into "white" and "black" sites, but in this case all of the sites in every other column are white sites. Third, a spin-flip two temperature model known as the Tome model, or North-East model, has also been studied [7]. In the Tome model, the spin-flip probability is calculated assuming that the bonds connecting the spin in question to the spins immediately to the right and immediately above are at one temperature, while the remaining two bonds are at another temperature.

For all of these spin-flip two temperature models, a continuous transition is found to exist, even away from equilibrium ($\beta_1 \neq \beta_2$), and the phase diagram looks qualitatively like Fig. 1. (Note that along the diagonal, $\beta_1 = \beta_2$, the models reduce to the equilibrium Ising model and the continuous transition occurs at the Onsagar point.) The static and dynamic universal behavior of the system along the entire critical line separating the disordered and ordered region remains the same as the equilibrium Ising model.

The situation is quite different for NESS Ising systems with spin-exchange dynamics. For example, the uniformly driven diffusive system [8], which consists of an Ising system with periodic boundary conditions, evolving subject to a uniform external field which biases the dynamics such that spin exchanges with + spins moving with (against) the field direction are enhanced (suppressed). This system has a continuous phase transition, but, unlike the spin-flip two temperature models discussed above, the universal behavior near the critical point differs from the equilibrium Ising model [9].

A related model is the randomly driven diffusive lattice gas [10], which is equivalent to the uniformly driven model except that the magnitude of the external field is not uniform, but instead is a zero mean, random variable in both space and time. The random drive stirs the system in the field direction, effectively changing the temperature of the spin exchanges in that direction. Thus, the randomly driven diffusive system is equivalent to a "strip" two temperature model [11], where spin exchanges in different directions occur at different temperatures. (See Fig. 2a.)

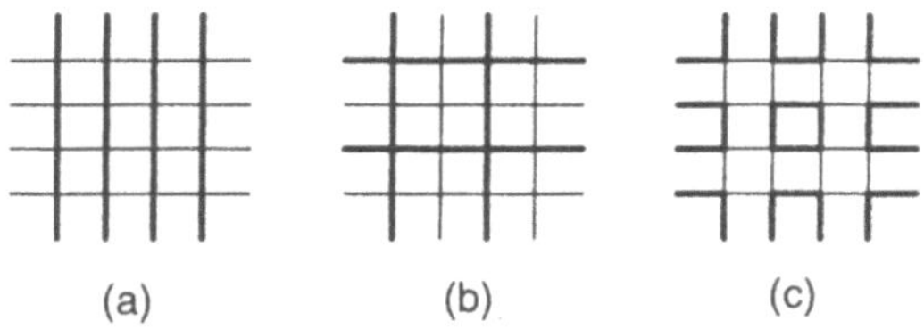

Fig. 2: Spin-exchange two temperature models. Spin exchanges at different bonds happen at one of two different temperatures. Thick lines and thin lines distinguish the temperature of the various bonds. a) the "strip" model, b) and c) "checkerboard" models.

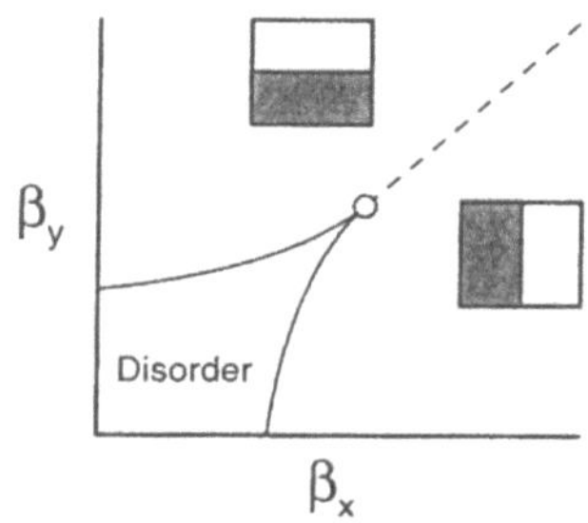

Fig. 3: Phase diagram for the strip spin-exchange two-temperature Ising model

The phase diagram [12] for this strip spin-exchange two-temperature Ising model is shown in Fig. 3. Along the $\beta_x = \beta_y$ diagonal the model reduces to the equilibrium Ising model. The critical point on this line again is the Onsager point. There are two distinct ordered phases, one with configurations where the interfaces separating the + spin rich from the + spin poor region is parallel to the y direction, and the other with configurations where those interfaces are parallel to the x direction. The phase boundaries separating the disordered phase from the ordered phases (indicated by solid lines in Fig. 3) are lines of continuous transitions. The critical behavior of the system near the non-equilibrium ordering transitions differs from that of the equilibrium Ising model [11, 13]. Interestingly, however, the critical properties of this model are believed to be in the universality class of another equilibrium system -- the uniaxial ferromagnetic with dipolar interactions [14].

The dashed line in Fig. 3, separating the two distinct ordered phases, is a line of first order transitions [12]. On the dashed line $\beta_x = \beta_y$, the system is in equilibrium, and the two ordered phases are degenerate.

The equilibrium Ising critical point is thus a *bicritical* point where two non-equilibrium critical lines meet a first order transition line separating two distinct ordered states. The shape of the critical lines near the bicritical point is a universal feature and is described by the equilibrium Ising-dipole crossover exponent ϕ, which is believed to be equal γ of the equilibrium Ising model.

Finally, I have also performed some preliminary simulations of two models we call "checkerboard" spin-exchange two temperature models. In these models spin exchanges on some bonds take place at one temperature ($=1/\beta_1$), while the rest take place at another temperature ($=1/\beta_2$). Figs. 2b and 2c indicate which bonds are at what temperature in these models. Note that there is no spatial anisotropy between the x and y directions introduced by the dynamics in these models. Although the simulations of these models are still in progress, we have determined that the phase diagrams for those models is qualitatively similar to those of the spin-flip two temperature models (Fig. 1), and initial indications are that this system remains in the universality class of the equilibrium Ising model along the entire critical line.

3. Summary and Discussion

Let us now return to the question posed in the introduction, and, in light of the above phenomenology, ask: What are the essential features of the dynamics required for a NESS Ising system to possess critical properties different from equilibrium Ising systems?

As one can see, many NESS Ising systems have critical properties that are in the equilibrium Ising universality class. Yet, some have critical properties that differ from that class. All of the spin-flip models mentioned have critical properties that are in the class of the equilibrium Ising model. This is true even for the strip and Tome models which have spatially anisotropic dynamics. On the other hand, both the uniformly driven diffusive system and the strip spin-exchange two temperature model have critical properties that differ from the equilibrium Ising model. However, the checkerboard spin-exchange two temperature models, which, similarly, have dynamics that conserve the magnetization, but which are spatially isotropic, have critical properties that are equivalent to the equilibrium Ising model.

Thus, it appears that two ingredients are necessary for NESS Ising systems to have ferromagnetic critical behavior that differs from the equilibrium Ising model. First, the nonequilibrium dynamics must conserve the magnetization, and, second, it must be spatial anisotropic. It will be interesting to see if these conclusions also hold for other NESS systems.

Acknowledgements

I thank Z. Racz, B. Schmittmann, and R.K.P. Zia for stimulating discussions. This research is supported in part by grants from the National Science Foundation through the Division of Materials Research and the Jeffress Memorial Trust.

References

[1] For a comprehensive review and further references, see: B. Schmittmann and R.K.P. Zia, to appear in **Phase Transitions and Critical Phenomena**, C. Domb and J.L. Lebowitz, editors, (Academic, N.Y.)

[2] See, e.g., D. J. Amit, **Field Theory, the Renormalization Group, and Critical Phenomena** (2nd edition: World Scientific, Singapore 1984).

[3] W. Lenz, Phys. Zeitschrift **21**, 613 (1920).

[4] R.J. Glauber, J. Math. Phys. **4**, 294 (1963).

[5] K. Kawasaki, Phys. Rev. **145**, 224 (1963).

[6] H.W.J. Blote, J.R. Heringa, A. Hoogland, and R.K.P. Zia, J. Phys. A: Math. Gen. **23**, 3799 (1990).

[7] H.W.J. Blote, J.R. Heringa, A. Hoogland, and R.K.P. Zia, Int. Journ. Mod. Phys. **B5**, 685 (1991).

[8] S. Katz, J.L. Lebowitz, and H. Spohn, Phys. Rev. **B28**, 1655 (1983) and J. Stat. Phys. **34**, 497 (1984).

[9] H.K. Janssen and B. Schmittmann, Z. Phys. **B64**, 503 (1986); K.-t.Leung and J.L.Cardy, J. Stat. Phys. **44**, 567 (1986), and 1087(1986); K. Gawedzki and A. Kupiainen, Nucl. Phys. **B2 69**, 45 (1986).

[10] P.L. Garrido, J.L. Lebowitz, C. Maes, and H. Spohn, Phys. Rev. **A42**, 1954 (1990); Z. Cheng, P.L. Garrido, J.L. Lebowitz, and J.L. Valles, Europhys. Lett. **14**, 507 (1991).

[11] B. Schmittmann and R.K.P. Zia, Phys. Rev. Lett. **66**, 357 (1991).

[12] K.E. Bassler and Z. Racz, preprint (1994).

[13] E.L.Praestgaard, H.Larsen, and R.K.P.Zia, Europhys. Lett. **25**, 447 (1994).

[14] B. Schmittmann, Europhys. Lett. **24** (2), 109 (1993).

Fourier-Accelerated Polymer Dynamics

B. Dünweg

Institut für Physik, Universität Mainz, Postfach 3980,
D-55099 Mainz, Germany

Fourier acceleration methods are applied to simulations of two–dimensional isolated ring polymers of up to $N = 64$ monomers. Three simulation schemes are compared: (i) a simple Langevin simulation with local updating, (ii) a Langevin algorithm with Fourier acceleration, and (iii) a Fourier accelerated Langevin algorithm combined with Metropolis acceptance of the moves (Force Biased Monte Carlo). In contrast to (i) and (ii), method (iii) is not hampered by systematic discretization errors, which, in case (ii), seem to grow systematically with chain length N. The results on the correlation time τ are not very accurate, however, the data are in rough agreement with $\tau \propto N^z$ with $z = 2.5$ (Rouse model, case (i)), $z \approx 0$ (ii) and $z \approx 1.5$ (iii). Taking into account that the simulation should produce correct statistics, method (iii) turns out to be the most efficient (except for $N = 8$). For longer chains, the speedup becomes substantial.

1 Introduction

The last decade has seen the development of a variety of sophisticated algorithms designed to "beat critical slowing down". Famous examples are cluster updating schemes like the Swendsen–Wang [1] and the Wolff [2] algorithm, or the Pivot method [3] for self–avoiding walks. All these methods use *nonlocal* updating schemes, i. e. a deliberately unphysical dynamics. Compared to local algorithms, they usually belong to different dynamic universality classes with a reduced dynamic exponent z (correlation time $\tau \propto R^z$, where R is some measure of the size of a correlated object like a critical cluster or a polymer chain).

Fourier acceleration [4, 5, 6] is one of these methods, but seems to be less well–known than cluster algorithms. It can be implemented rather straightforwardly for a wide variety of systems, but tends to be less efficient than other "smart" schemes. However, if no better method is at hand, it may still be helpful compared to simple local schemes. The philosophy of the algorithm is based on the notion of Fourier modes which relax more and more slowly as the wavelength increases. Hence the algorithm "kicks" the long–wavelength modes particularly strongly. In general, these modes are not eigenmodes of the system, but nevertheless the increase of the relaxation time with wavelength can be partly canceled out. Some orientation on how to do the "kicking" usually is obtained from trivial noninteracting limiting cases: In lattice field theory, this is the "free field theory", and it is rather obvious that the corresponding trivial polymer system must be the random walk chain with Rouse dynamics [7], in which the nontrivial excluded–volume interaction is switched off. In such a noninteracting case the relaxation time can be calculated analytically as a function of wavelength.

Springer Proceedings in Physics, Vol. 78
Computer Simulation Studies in Condensed-Matter Physics VII
Editors: D.P. Landau · K.K. Mon · H.-B. Schüttler

The "acceleration factor" for a given mode is then chosen proportional to its estimated relaxation time (for more details, see below). The better one guesses the behavior of the real system from the noninteracting case, the more efficiently critical slowing down will be canceled out. Hence, in some cases it may be more advisable to choose the acceleration factors on the basis of empirical measurements of the decay of modes, or of known scaling laws.

The scheme starts from a representation in real space in order to calculate the interactions, transforms into Fourier space to do the update and transforms the updated configuration back into real space. Since two transformations are involved every single timestep, system sizes which are a power of two are by far the most efficient because in this case Fast Fourier Transform (FFT) algorithms [8] work best. The method, as it has been used so far, is based on collective Langevin-type updates, and hence restricted to *continuous* degrees of freedom. This also means that one has to cope with the typical discretization errors [9] and numerical instabilities [4, 10] of Langevin schemes. Both problems can be controlled by combining the Langevin updating with a Metropolis acceptance criterion (Force Biased Monte Carlo) [10], which will be discussed at the end.

In what follows I report first attempts to apply Fourier acceleration methods to complex fluid problems, in particular polymers, which, to my knowledge, has not been done yet. As a first toy model a single two-dimensional ring polymer is studied. This is easy to simulate, and the involved physics is well understood. In two dimensions, one can interpret the x and y coordinates of a particle as a single complex number, which permits to perform one single complex FFT. A ring polymer (in contrast to a linear chain) was chosen because its strict translational invariance along the chain makes the solution of the random walk Rouse dynamics (to estimate the acceleration factors) particularly easy[1]. The results for the single ring look rather promising, but it remains to be seen yet if the method will prove useful for really interesting problems like multichain systems or tethered membranes.

2 Model, Theory and Simulation Method

The configuration is described by the monomer coordinates $\vec{r}_n$, $n = 0, \ldots, N-1$. A purely repulsive cosine potential between all monomers models the excluded-volume interaction:

$$U_{EV}(r) = \begin{cases} \frac{\varepsilon}{2}\left[1 + \cos\left(\frac{\pi r}{\sigma}\right)\right] & r \leq \sigma \\ 0 & r \geq \sigma. \end{cases} \tag{1}$$

Originally, a purely repulsive Lennard-Jones potential was implemented. However, the Langevin simulations based on this potential very easily ran into numerical instabilities, and hence this approach was discarded. The cosine potential permits a timestep which is at least an order of magnitude larger. In what follows, the length and energy units will be fixed by setting $\varepsilon = \sigma = 1$. Simulations were then run at a temperature $T = 0.2$ (Boltzmann's constant is also set to unity). The monomers n and $n+1$, $n = 0, \ldots, N-2$,

[1] First tests indicate that the modes and acceleration factors derived below for a ring polymer do not yield significant speedup for a linear chain. Apparently, a linear chain needs different modes. It is not yet clear if these can be implemented efficiently as well.

as well as 0 and $N-1$ were connected via harmonic springs, $U_h(r) = (k/2)r^2$, with k chosen as 1.

In the Fokker–Planck picture, any Langevin dynamics in which only the $\vec{r}_n$ occur as stochastic variables (i. e. no momenta) can be described via the equation of motion [11]

$$\frac{\partial}{\partial t} P(\{\vec{r}_i\}, t | \{\vec{r}_i^0\}, 0) = \sum_{mn} \frac{\partial}{\partial \vec{r}_m} \cdot \overleftrightarrow{D}_{mn} \cdot \left(\frac{\partial}{\partial \vec{r}_n} - \beta \vec{F}_n \right) P(\{\vec{r}_i\}, t | \{\vec{r}_i^0\}, 0). \qquad (2)$$

Here $P(\{\vec{r}_i\}, t | \{\vec{r}_i^0\}, 0)$ is the conditional probability density of a transition from configuration $\{\vec{r}_i^0\}$ at time 0 to $\{\vec{r}_i\}$ at time t. $\vec{F}_n$ are the forces derived from the potential given above, while $\beta = 1/T$. $\overleftrightarrow{D}_{mn}$ is the diffusion tensor, which in general couples monomer m with monomer n. In the most general case, $\overleftrightarrow{D}_{mn}$ can depend on the configuration, but here we will only consider constant tensors.

At the heart of Fourier acceleration lies the observation that *the equilibrium solution of Eqn. 2, i. e. the canonical distribution governed by the potential energy, does not depend on the choice of* $\overleftrightarrow{D}_{mn}$ *at all.* Hence, one may try to choose it in a clever way in order to speed up the algorithm (however, it has to be symmetric and positive-definite). This situation is similar to "smart" Monte Carlo algorithms which exploit the freedom of choice of the transition probabilities.

In the simplest Langevin scheme, Eqn. 2 is translated to a discretized algorithm using a finite timestep h [11, 12]:

$$\vec{r}_m(t+h) = \vec{r}_m(t) + h \sum_n \overleftrightarrow{D}_{mn} \, \beta \vec{F}_n + \sqrt{2h} \vec{\rho}_m, \qquad (3)$$

where the $\vec{\rho}_m$ are random displacements, whose means vanish, while $\langle \vec{\rho}_m \otimes \vec{\rho}_n \rangle = \overleftrightarrow{D}_{mn}$. The details of the distribution function of the $\vec{\rho}_m$ do not matter (as long as all moments exist) and one can hence use a uniform distribution [12].

In order to find a good choice for $\overleftrightarrow{D}_{mn}$, one solves Eqn. 2 for the random walk Rouse model [7]: In this case, only the harmonic springs remain in the interaction. Since Rouse dynamics is based on uncorrelated random displacements, $\overleftrightarrow{D}_{mn} = D_0 \delta_{mn} \overleftrightarrow{1}$ in the non-accelerated case. In what follows, we will set $D_0 = 1$ in order to fix the time units. Rouse modes are introduced via $\vec{X}_p = \sum_{n=0}^{N-1} F_{pn} \vec{r}_n$, $p = 0, \ldots, N-1$. Here the matrix F_{pn} shall be defined in a slightly different way than usually done [7]:

$$F_{pn} = \sqrt{\frac{2}{N}} \cos\left(\frac{2\pi pn}{N} + \frac{\pi}{4} \right). \qquad (4)$$

By this definition, F_{pn} is real, symmetric and orthogonal, and hence also $\vec{r}_n = \sum_{p=0}^{N-1} F_{pn} \vec{X}_p$. This definition also means that the slow modes are not only those close to $p = 0$, but also to $p = N$: After some algebra, Eqn. 2 is transformed to the new variables $\vec{X}_p$, whose relaxation times turn out to be proportional to $a(p) = 2\left[1 - \cos(2\pi p/N)\right]^{-1}$. This is then chosen as acceleration factor, i. e. in the Fourier accelerated scheme the diffusion tensor is chosen as $\overleftrightarrow{D}_{mn} = \sum_p F_{mp} a(p) F_{np} \overleftrightarrow{1}$. A little care must be taken of the center-of-mass motion corresponding to $p = 0$, since $a(p) \to \infty$ for $p \to 0$. Since the center-of-mass position is irrelevant for the chain structure, we use $a(0) = 1$ by arbitrary choice ($a(0) = 0$ would violate the condition of positive–definiteness). $\vec{\rho}_n$ is constructed from random displacements $\vec{\xi}_p$ in Fourier space via $\vec{\rho}_n = \sum_p F_{np}\sqrt{a(p)}\vec{\xi}_p$ with $\langle \vec{\xi}_p \rangle = 0$ and $\langle \vec{\xi}_p \otimes \vec{\xi}_q \rangle = \delta_{pq} \overleftrightarrow{1}$.

The FFT was done using a library routine from Ref. [8]. For the parameters mentioned above, a timestep $h = 10^{-3}$ was used without encountering instabilities.

3 Simple vs. Fourier Accelerated Langevin

For two–dimensional self–avoiding walks, the asymptotic relation $\langle \vec{R}^2 \rangle \propto N^{2\nu}$ with $\vec{R} = \vec{r}_{n+N/2} - \vec{r}_n$ and $\nu = 3/4$ holds exactly [13], and was used to test the correctness of the algorithms. The autocorrelation time τ was estimated by a time integral over the normalized autocorrelation function $\langle \vec{R}(t) \cdot \vec{R}(0) \rangle / \langle \vec{R}^2 \rangle$. For local dynamics, one expects the Rouse result $\tau \propto N^{1+2\nu}$ [7]. For comparison, both the Fourier accelerated algorithm as well as local updating (i. e. $a(p) = 1$, where the FFTs were removed from the program) were employed. Data about the performance are summarized in Table 1. One sees that the correlation time of the local algorithm increases dramatically with chain length (in rather good agreement with the Rouse scaling law), while the Fourier acceleration basically keeps τ constant. However, in order to really measure the dynamic critical exponent of the Fourier algorithm, one would need significantly longer chains and much better statistics.

$\langle \vec{R}^2 \rangle$ as a function of chain length is displayed in Fig. 1. The data are, at first glance, somewhat surprising: The Fourier accelerated algorithm systematically measures a larger $\langle \vec{R}^2 \rangle$ than the local scheme, the latter being in better agreement with the theoretical power law. The explanation is that Langevin simulations are always subject to discretization errors which grow with the timestep. For sufficiently small h, the error in $\langle \vec{R}^2 \rangle$ should be proportional to h [9]. However, the accelerated algorithm effectively has a *much* bigger timestep than the local algorithm, due to the acceleration factors. Since $a(1)$ grows with N, this effect gets more and more pronounced the larger the system gets. This seems to be a general weakness of the Fourier accelerated Langevin scheme: As a test, the $N = 32$ ring was also run at $h = 10^{-4}$. $\langle \vec{R}^2 \rangle$ decreased, as expected, but only slightly. Apparently, a timestep which works rather close to the stability limit (as $h = 10^{-3}$ does), introduces an an enormous discretization error which is so large that it is no longer in the linear regime. Of course, a reduction of h by orders of magnitude is expected to cure the problem, but then the acceleration scheme is no longer worthwhile. Instead, it is easier to generalize the method to Force Biased Monte Carlo, similar to the procedure adopted in Ref. [10].

Table 1: Summary of performance results for pure Langevin simulations of ring polymers. For meaning of symbols see text. Errors in τ were roughly estimated. The CPU times refer to an IBM RS 6000 model 250 RISC workstation.

	without FFT			with FFT		
N	# steps	CPU time / step	τ	# steps	CPU time / step	τ
8	10^7	6.1×10^{-5} sec.	4.5 ± 0.5	10^6	1.4×10^{-4} sec.	0.8 ± 0.2
16	10^7	1.5×10^{-4} sec.	35 ± 5	10^6	2.8×10^{-4} sec.	0.7 ± 0.3
32	10^8	4.0×10^{-4} sec.	170 ± 30	10^6	6.4×10^{-4} sec.	1 ± 0.2
64	7.636×10^7	1.2×10^{-3} sec.	1000 ± 100	10^6	1.8×10^{-3} sec.	0.9 ± 0.1

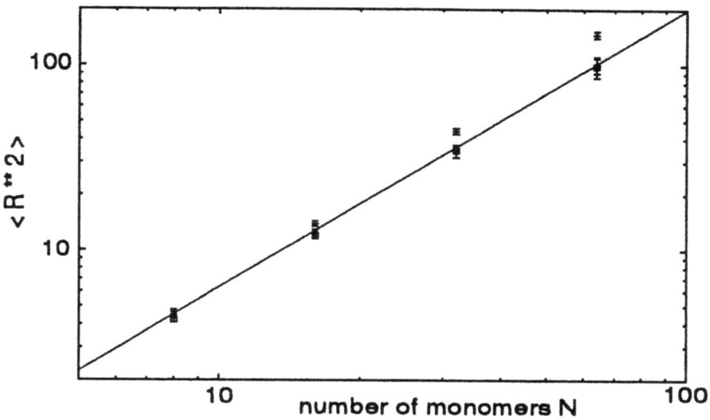

Figure 1: The ring polymer analogue of the end–end distance, $\left\langle \vec{R}^2 \right\rangle$ (see text) as a function of chain length N. Diamonds correspond to a pure Langevin simulation without acceleration, crosses to Fourier accelerated Langevin dynamics, and squares to Fourier accelerated Force Biased Monte Carlo. Note that the error bars only take the *statistical* error into account, disregarding the systematic discretization error of the pure Langevin schemes. The straight line is drawn as guide to the eye, displaying the slope of the theoretical $N^{1.5}$ power law.

4 Fourier Accelerated Force Biased Monte Carlo

This method just reinterprets the move generated by the Langevin scheme as a Monte Carlo trial move which can be accepted or rejected. Detailed Balance is satisfied if the ratio of the probability of a given move to the probability of its inverse move is the Boltzmann factor. However, the probability of a move is given by the *a priori* probability times the probability of acceptance. While in the *simple* Metropolis scheme the *a priori* probabilities are the same for the move as well as its inverse, this is not the case if they are generated by the Langevin scheme. However, these probabilities can be calculated and enter the acceptance criterion. This procedure automatically generates a canonical ensemble, thereby *totally removing the discretization error* and stabilizing the algorithm. In the present case, one has to write down the *a priori* probabilities in Fourier space and to use Gaussian random numbers for the Langevin displacements (ratios of Gaussian functions are easy to calculate). For further details see Ref. [10]. Apart from the additional Metropolis check, all run parameters were left unchanged.

The results on $\left\langle \vec{R}^2 \right\rangle$ are included in Fig. 1. As expected, the data are now in agreement with the correct power law. Data about the performance are summarized in Table 2. In contrast to the pure Langevin scheme, one now observes a pronounced increase of τ with N. This is due to the increased rejection rate at larger N which is necessary to correct for the larger and larger discretization error. Nevertheless, τ seems to increase slowlier with N than for the local scheme ($\tau \propto N^z$ with z roughly around 1.5). Taking into account both τ as well as the CPU time per step, one finds that the method is more efficient than the local scheme except for the shortest chain. For $N = 64$, the speedup is already a factor of roughly 20.

Table 2: Same as Table 1 for the Fourier accelerated Force Biased Monte Carlo scheme.

N	# steps	CPU time / step	τ
8	10^6	2.4×10^{-4} sec.	1.3 ± 0.3
16	10^6	4.8×10^{-4} sec.	2.7 ± 0.3
32	10^6	1.1×10^{-3} sec.	10 ± 4
64	10^6	2.5×10^{-3} sec.	25 ± 15

References

[1] R. H. Swendsen and J. S. Wang, Phys. Rev. Lett. **58**, 86 (1987).

[2] U. Wolff, Phys. Rev. Lett. **62**, 361 (1989).

[3] N. Madras and A. D. Sokal, J. Stat. Phys. **50**, 109 (1988).

[4] G. G. Batrouni *et al.*, Phys. Rev. D **32**, 2736 (1985).

[5] A. L. Scheinine, Phys. Rev. B **39**, 9368 (1989).

[6] I. T. Drummond and R. R. Horgan, Physics Letters B **302**, 271 (1993).

[7] M. Doi and S. F. Edwards, *The Theory of Polymer Dynamics* (Clarendon Press, Oxford, 1986).

[8] W. H. Press, B. P. Flannery, S. A. Teukolsky, and W. T. Vetterling, *Numerical Recipes in C* (Cambridge University Press, Cambridge, 1988).

[9] A. Greiner, W. Strittmatter, and J. Honerkamp, J. Stat. Phys. **51**, 95 (1988).

[10] D. Ceperley, M. H. Kalos, and J. L. Lebowitz, Macromolecules **14**, 1472 (1981).

[11] H. Risken, *The Fokker–Planck Equation* (Springer–Verlag, Berlin, 1984).

[12] B. Dünweg and W. Paul, Int. Journ. Mod. Phys. C (Physics and Computers) **2**, 817 (1991).

[13] B. Nienhus, in *Phase Transitions and Critical Phenomena*, edited by C. Domb and J. L. Lebowitz (Academic Press, London, 1987), Vol. 11.

A Multicanonical Study of Non-Polar Amino Acids

U.H.E. Hansmann[1] and *Y. Okamoto*[2]

[1]Supercomputer Computations Research Institute (SCRI), The Florida State University, Tallahassee, FL 32306, USA

[2]Department of Physics, Nara Women's University, Nara 630, Japan

We perform a multicanonical simulation of non polar amino acids and study their α-helix propensities. Starting simulations from completely random initial conformations, we have calculated various quantities like the average % helix, free energy differences of helix-coil transitions, and the Zimm-Bragg s parameters over a wide range of temperatures. The results are shown to be in agreement with recent experimental results.

Prediction of three-dimensional structure of proteins solely from their amino acid sequence remains one of the unsolved problems in bioscience. The difficulty with the conventional methods such as molecular dynamics lies in the fact that simulations strongly depend on the initial conditions and cannot reach thermal equilibrium within the presently available computer resources. The *multicanonical algorithm*[1] is known to alleviate the above difficulty in other thermodynamic systems with frustration such as spin glasses.[2] The application of this algorithm to protein folding problem was recently proposed.[3] Here we report multicanonical simulations on α-helix propensities of homo-oligomers of nonpolar amino acids.[4] Recent experimental measurements [5, 6] suggest large differences in helix propensities among the amino acids while the older host-guest method[7] indicated small differences (for a review see Ref. [5]). Our aim is to reproduce this experimental results in a numerical simulation for three characteristic amino acids: alanine (helix former), glycine (helix breaker), and valine (helix indifferent). The same homo-oligomers were studied by Monte Carlo simulated annealing[8] in Ref. [9]. However, that investigation suffers from the problem that due to finite annealing steps simulated annealing is not well suited for a reliable calculation of thermodynamic quantities. On the other hand, with the multicanonical algorithm one can obtain any thermodynamic quantity for a wide range of temperatures from only *one* production run.[2, 3]

The multicanonical ansatz consists of three steps. First multicanonical weight factors are constructed in a recursive way.[2] This allows to simulate a "multicanonical" ensemble[1] in which all energies enter with equal probability. With respect to this ensemble equilibrium configurations are generated by the standard Monte Carlo procedure. Since the energy is forced onto a $1d$ random walk by performing a simulation in this new ensemble, one avoids getting trapped in a local minimum and the probability of finding the global minimum is increased. In the last step canonical expectation values are calculated by re-weighting[10] over a wide range of temperatures.[2, 3]

We considered three homo-oligomers of 10 amino acids, $(Ala)_{10}$, $(Val)_{10}$, and $(Gly)_{10}$. The computer code KONF90 [11] was modified to acomodate the multicanonical method. The semi-empirical potential energy function that we used is given by the sum of the electrostatic term E_{es}, 12-6 Lennard-Jones term E_{LJ}, and hydrogen-bond term E_{hb} for all

Springer Proceedings in Physics, Vol. 78
Computer Simulation Studies in Condensed-Matter Physics VII
Editors: D.P. Landau · K.K. Mon · H.-B. Schüttler

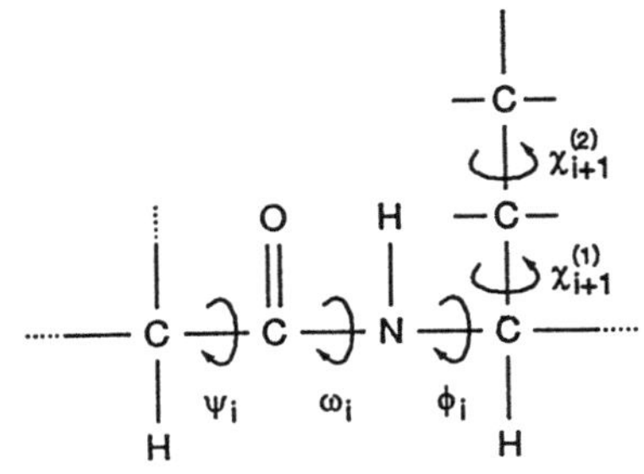

Figure 1: Definition of the dihedral angles.

pairs of atoms in the peptide together with the torsion term E_{tors} for all torsion angles:

$$E_{tot} = E_{es} + E_{LJ} + E_{hb} + E_{tors} \quad , \tag{1}$$

$$E_{es} = \sum_{(i,j)} \frac{332 q_i q_j}{\epsilon r_{ij}}, \tag{2}$$

$$E_{LJ} = \sum_{(i,j)} \left(\frac{A_{ij}}{r_{ij}^{12}} - \frac{B_{ij}}{r_{ij}^{6}} \right), \tag{3}$$

$$E_{hb} = \sum_{(i,j)} \left(\frac{C_{ij}}{r_{ij}^{12}} - \frac{D_{ij}}{r_{ij}^{10}} \right), \tag{4}$$

$$E_{tors} = \sum_{l} U_l \left(1 \pm \cos(n_l \alpha_l)\right). \tag{5}$$

r_{ij} is the distance between the atoms i and j, and α_l is the torsion angle for the chemical bond l. The parameters ($q_i, A_{ij}, B_{ij}, C_{ij}, D_{ij}, U_l$ and n_l) for the energy function were adopted from ECEPP/2 as documented in Ref. [12]. For a definition of the dihedral angles ϕ, ψ, χ and ω, which represent the true degrees of freedom see Fig. 1. The peptide-bond dihedral angles ω were fixed at the value 180° and the dielectric constant was set to 2. Because one can avoid the complications of electrostatic and hydrogen-bond interactions of side chains with the solvent for nonpolar amino acids, explicit solvent molecules were neglected. One Monte Carlo step consists of updating all the torsion angles in the backbone and side chains. For each homo-oligomer, several preliminary runs with 10^4 Monte Carlo steps were made to determine the multicanonical weight factors, and then one production run with 10^5 Monte Carlo steps was made from a completely random initial conformation.

The criterion we adopt for α-helix formation is as follows: We consider that a residue is in the α-helix state when the dihedral angles (ϕ, ψ) fall in the range $(-70\pm20°, -37\pm20°)$. The length ℓ of a helical segment is then defined by the number of successive residues which are in the α-helix state. The number n of helical residues in a conformation is defined by the sum of ℓ over all helical segments in the conformation. In Fig. 2 we show the average % helix per residue $\frac{<n>}{N}$ ($N = 10$) as a function of temperature for each homo-oligomer. $(\mathrm{Ala})_{10}$ is a strong helix former with % helix varying from $\sim$ 80 % at $T = 200$ K to $\sim$ 50 % at $T = 400$ K, and $(\mathrm{Gly})_{10}$ is a strong helix breaker with % helix varying from $\sim$ 10 % at $T = 200$ K to $\sim$ 7 % at $T = 400$ K, while $(\mathrm{Val})_{10}$ comes in between the two with % helix varying from $\sim$ 35 % at $T = 200$ K to $\sim$ 17 % at $T = 400$ K. This is in accord with the experimental results. [5, 6] From the average of n and ℓ one can calculate the helix propagation parameter s of the Zimm-Bragg model [13]

$$s = \frac{(< l > -1)(1 - \frac{<n>}{N})}{< l > (1 - \frac{<n>}{N}) - \frac{<n>}{N}} \, . \tag{6}$$

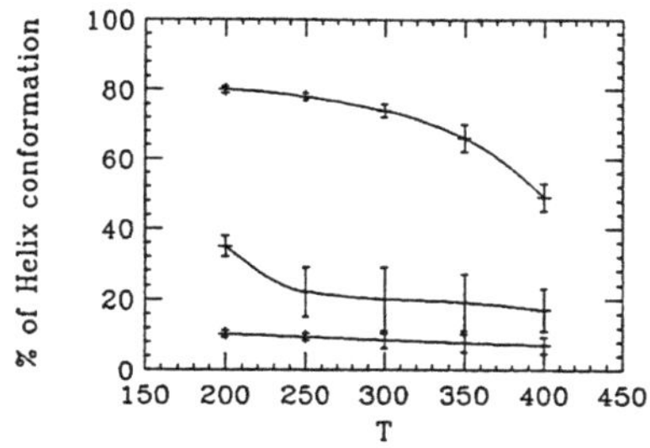

Figure 2: Average % helix per residue $\frac{<n>}{N}$ ($N = 10$) as a function of temperature for the three homo-oligomers, $(Ala)_{10}$, $(Val)_{10}$, and $(Gly)_{10}$.

Table 1: Free energy differences ΔG, enthalpy differences ΔH, and entropy differences $T\Delta S$ (in kcal/mol) between helix and non-helix states.

Peptide	T	ΔG	ΔH	$T\Delta S$
$(Ala)_{10}$	250	-4.3(1.0)	-10.6(1.1)	-6.5(1.5)
	300	-3.0(8)	-10.1(2.5)	-7.2(2.6)
	350	-1.9(5)	-10.1(2.8)	-8.3(2.9)
$(Val)_{10}$	250	0.41(28)	-2.1(1.8)	-1.0(1.8)
	300	0.79(54)	-0.94(53)	-1.6(8)
	350	1.1(5)	-1.8(1.3)	-3.8(1.1)
$(Gly)_{10}$	250	2.7(1.2)	1.3(2.1)	-0.4(2.4)
	300	3.1(1.0)	0.28(1.7)	-1.9(2.0)
	350	3.6(9)	-0.29(2.0)	-4.1(2.2)

This parameter was also obtained by experiments.[5, 6] We found $s(\text{Ala}) = 1.5 \sim 1.6$, $s(\text{Val}) = 0.37 \sim 0.45$, and $s(\text{Gly}) = 0.13 \sim 0.16$ around the experimentally relevant temperature ($\sim 0°$ C). These values are in remarkable agreement with the experiments,[6] where they give $s(\text{Ala}) = 1.99 \sim 2.19$, $s(\text{Val}) = 0.20 \sim 0.93$, and $s(\text{Gly}) = 0.02 \sim 0.57$.

In Table 1 we present the free energy differences $\Delta G \equiv G_H - G_C$, enthalpy differences ΔH, and entropy differences $T\Delta S$ between helix (H) and non-helix (C) states. Here, a conformation is considered as in the helix state if it has a segment with helix length $\ell \geq 3$. Note that $\ell = 3$ corresponds to roughly one turn of the α-helix . The free energy differences were calculated from $\Delta G = -RT \ln \frac{N_H}{N_C}$, where N_H and N_C are average numbers of conformations in helix and non-helix states, respectively. The enthalpy differences were obtained from $\Delta H = E_H - E_C$, where E_H and E_C are average potential energies in helix and non-helix states, respectively. Finally, the entropy differences were derived from ΔG and ΔH by the relation $T\Delta S = \Delta H - \Delta G$. It is clear from the table that around temperatures near 0° C $(Ala)_{10}$ favors helix state with $\Delta G = -3 \sim -4$ kcal/mol and $(Gly)_{10}$ favors non-helix state with $\Delta G = 2.7 \sim 3$ kcal/mol, while $(Val)_{10}$ slightly favors non-helix state with $\Delta G = 0.4 \sim 0.8$ kcal/mol. These results again support the experimental fact that Alanine is a helix former and Glycine is a helix breaker, while Valine comes in between the two. Note that for each homo-oligomer the entropy contribution $-T\Delta S$ monotonically increases with temperature as it should because of the increased thermal fluctuations. Note also that ΔH is large negative for $(Ala)_{10}$, whereas it is small for $(Val)_{10}$ and $(Gly)_{10}$, suggesting that ΔH is a key factor for helix stability. This can be

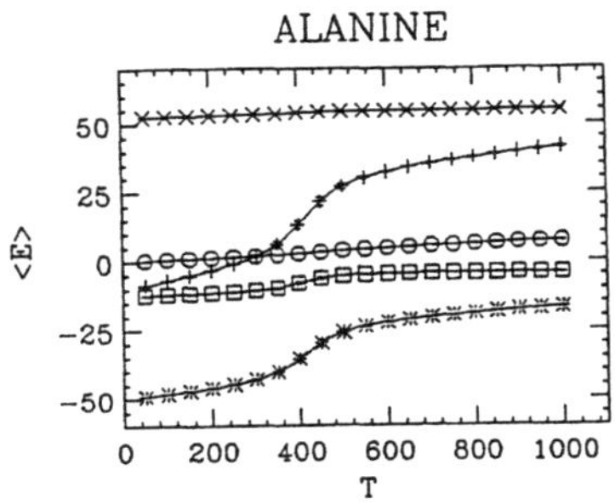

Figure 3: Total Energy E_{tot} (+),Coulomb term $E_{es}(x)$,Lennard-Jones term $E_{LJ}(*)$, Hydrogen-bond term $E_{HB}(\square)$ and Torsion energy $E_{tors}(o)$ as a function of temperature.

understood by the fact that helical conformations is one of the ideal conformations that minimize the Lennard-Jones term E_{LJ}. Fig. 3 shows for Alanine that the behavior of the energy as a function of temperature is indeed dominated by this term. The relative dominance of the Lennard-Jones energy depends on the geometry of the side chains. However, we found for all three homo-oligomers studied in the present work that the lowest energy configuration is helical.

To conclude let us remark that we observe direct *folding* of helices from completely *random* initial conformations. Our numerical results could qualitatively and quantitatively reproduce recent experimental results. We have demonstrated the superiority of multicanonical algorithm to conventional methods by calculating various thermodynamic quantities as functions of temperature from only *one* simulation run, which was not possible previously.

Acknowledgements:

This work was supported by a Grant-in-Aid for Scientific Research from the Japanese Ministry of Education, Science and Culture, by the U.S. Department of Energy, contract DE-FC05-85ER250000 and by MK Industries. Our simulations were performed on the cluster of RISC workstations at SCRI and the HITAC S820/80 at NLMS, Okazaki, Japan.

References

[1] Berg, B.A. & Neuhaus, T. *Phys. Lett.* **B267**, 249–253 (1991); Berg, B.A. & Neuhaus, T. *Phys. Rev. Lett.* **68**, 9–12 (1992).

[2] Berg, B.A. & Celik, T. *Phys. Rev. Lett.* **69**, 2292–2295 (1992); Berg, B.A., Celik, T. & Hansmann, U. *Europhysics Lett.* **22**, 63–68 (1993); Celik, T., Hansmann, U. & Katoot, M. *J. Stat. Phys.* **73** (1993) 775-788.

[3] Hansmann, U. & Okamoto, Y. *J. Comp. Chem.* **14**, 1333–1338 (1993).

[4] Okamoto, Y. & Hansmann, U.,*Predicting α-helix propensities of nonpolar amino acids by multicanonical algorithm*, preprint NWU 1/94, submitted for publication.

[5] Chakrabartty, A. & Baldwin, R.L. in *Protein Folding: In Vivo and In Vitro* (eds Cleland, J. & King, J.) 166–177 (ACS Press, Washington, D.C., 1993).

[6] Padamanabhan, S., Marqusee, S., Ridgeway, T., Laue, T.M. & Baldwin, R.L. *Nature* **344**, 268–270 (1990); Lyu, P.C., Liff, M.I., Marky, L.A. & Kallenbach, N.R. *Science* **250**, 669–673 (1990); Chakrabartty, A., Schellman, J.A. & Baldwin, R.L. *Nature* **351**, 586–588 (1991); Gans, P.J., Lyu, P.C., Manning, M.C., Woody, R.W. & Kallenback, N.R. *Biopolymers* **31**, 1605-1614 (1991); Stellwagen, E., Park, S.-H., Shalango, W. & Jain, A. *Biopolymers* **32**, 1193–1200 (1992).

[7] Sueki, M.,Lee, S.,Power, S.P.,Denton, J.B.,Konishi, Y. and Scheraga, H.A. *Macromolecules* **17**, 148-155 (1984).

[8] Kirkpatrick, S., Gelatt, C.D., Jr. & Vecchi, M.P. *Science* **220**, 671–680 (1983).

[9] Okamoto, Y. *Proteins struct. funct. Genet.* (in the press).

[10] A.M. Ferrenberg and R.H. Swendsen, *Phys. Rev. Lett.*, **61**, 2635 (1988); **63** , 1658(E) (1989), and references given in the erratum.

[11] Kawai, H., Okamoto, Y., Fukugita, M., Nakazawa, T. & Kikuchi, T. *Chem. Lett.* **1991**, 213–216 (1991); Okamoto, Y., Fukugita, M., Nakazawa, T. & Kawai, H. *Protein Engng* **4**, 639–647 (1991).

[12] Momany, F.A., McGuire, R.F., Burgess, A.W. & Scheraga, H.A. *J. Phys. Chem.* **79**, 2361–2381 (1975); Némethy, G., Pottle, M.S. & Scheraga, H.A. *J. Phys. Chem.* **87**, 1883–1887 (1983); Sipple, M.J., Némethy, G. & Scheraga, H.A. *J. Phys. Chem.* **88**, 6231–6233 (1984).

[13] Zimm, B.H. & Bragg, J.K. *J. Chem. Phys.* **31**, 526–535 (1959).

Classical and Quantum Two-Dimensional Fluids in the Gibbs Ensemble

F. Schneider, M.-O. Ihm, and P. Nielaba

Institut für Physik, KoMa 331, Universität Mainz, D-55099 Mainz, Germany

Abstract. We study the properties of model fluids in two spatial dimensions with Gibbs ensemble Monte Carlo (GEMC) techniques. In particular in the first part of the paper we study the entropy driven phase separation in case of a nonadditive symmetric hard disc fluid and locate by a combination of GEMC with finite size scaling techniques the critical line of nonadditivities as a function of the system density, which separates the mixing/demixing regions, we compare with a simple approximation. In the second part we successfully combine path integral Monte Carlo (PIMC) and GEMC techniques in order to locate the gas–liquid coexistence densities for a fluid with classical degrees of freedom and internal quantum states, a comparison with NVT-ensemble results and mean field predictions is presented.

1. Phase Transitions in Nonadditive Symmetric Hard Disc Fluids

Phase transitions in systems with purely repulsive interaction have got much attention recently [1]. In this section we consider a system of hard discs (of diameter d) with N_A particles of type A and N_B particles of type B and interaction potential:

$$U(r_{12})=\begin{cases} 0 & : \quad r_{12} > d_{s_1 s_2} \\ \infty & : \quad r_{12} < d_{s_1 s_2} \end{cases} \tag{1}$$

r_{12} is the distance of two particles, $s_1, s_2 \in \{A, B\}$ are their species and $d_{AA} = d_{BB} = d$, $d_{AB} = d + \Delta/2$. The total number of particles N and the total volume V is fixed and thus the average density $\rho^* = \rho * d^2 = Nd^2/V$. Due to the additional repulsion between A and B- type particles we expect a phase separation into a A-rich and a B-rich fluid phase for large values of $\Delta > \Delta_c$ and fixed total density. For $\Delta = 0$ we have a pure hard disc fluid with no phase separation in the fluid phase. In order to locate the critical values Δ_c as a function of ρ^* we perform GEMC simulations [2]. Since the phase separation is driven by entropy we expect only a small interfacial free energy in case of phase coexistence. In order to be able to locate the critical points even in such extreme situations we combine the GEMC with the block analysis finite size scaling techniques [3]. The GEMC results [4] were obtained with

Springer Proceedings in Physics, Vol. 78
Computer Simulation Studies in Condensed-Matter Physics VII
Editors: D.P. Landau · K.K. Mon · H.-B. Schüttler

Table 1: *Critical line for nonadditive hard discs obtained by the cumulant intersection method. For the largest values of Δ_c the critical density ρ_c^* was found by density variation, the smallest critical values of Δ_c were obtained for fixed density.*

ρ_c^*	Δ_c/d
0.6	0.562 ± 0.04
0.55	0.66 ± 0.01
0.5	0.789 ± 0.01
0.499 ± 0.01	0.85
0.5 ± 0.02	0.9
0.508 ± 0.03	0.95
0.46 ± 0.02	1

N=512 particles and about 10^6 Monte Carlo steps, where each step consisted of 400 attempted moves, 20 particle exchange and 2 volume change attempts. Cell ocuppancy lists were successfully used to speed up the procedure, the overall cumputing time for was about 4300 CPU hours on RISC 6000/250 workstations.

In order to get a rough idea about the critical line l_c, defined by $\rho_c^*(\Delta_c)$ we computed l_c in analogy to a study in three dimensions [5] by convex envelope arguments for the free energy and arrive at the compact expression:

$$\rho_c^*(\Delta_c) = \frac{4}{\pi}\left(1 - \sqrt{\frac{\Delta}{d+\Delta}}\right). \tag{2}$$

The critical points were obtained by inspection of number-difference histograms $P_L(N_A - N_B)$ on different length scales L obtained by subdivision of the simulation boxes of sizes V_1 and V_2 ($V_1 + V_2 = V$) into smaller subsystems of size $L \times L$. For $\Delta < \Delta_c$ the distributions are all singly peaked, for larger Δ we obtain a single peak structure of $P_L(N_A - N_B)$ for large L and a double peak structure for small L. An analysis of these histgrams with the cumulants $U_L = 1 - < (N_A - N_B)^4 >_L /3 < (N_A - N_B)^2 >_L^2$ allows a determination of critical points, due to the cumulants L- invariance at the critical point. The GEMC results for the critical line are presented in table 1. Similarily to the prediction of eq.(2) ρ_c is a decreasing function of Δ_c, however at a given density the GEMC results for Δ_c are about 20% larger than the predictions of eq.(2).

In total by a combination of GEMC and finite size scaling techniques we obtain results for the critical line for nonadditive symmetric hard discs even in cases where the interfacial free energy is purely entropic. Further details will be presented elsewhere, future work for asymmetric hard discs is in preparation.

2. Path Integral Monte Carlo Simulations in the Gibbs Ensemble

In this section we present results of a novel combination of GEMC and path integral Monte Carlo simulation techniques. In particular we study the gas–liquid transition of a model fluid with internal quantum states. The Hamiltonian is given by:

$$H = \sum_{i=1}^{N} \frac{\mathbf{p}_i^2}{2M} - \frac{\omega_0}{2} \sum_{i=1}^{N} \sigma_i^x + \sum_{i<j} U(r_{ij}) - \sum_{i<j} J(r_{ij}) \sigma_i^z \sigma_j^z \quad . \tag{3}$$

M is the particle mass, $\mathbf{p}_i$ is the momentum of particle i, r_{ij} is the distance between particle i and j, σ^x and σ^z are the usual Pauli spin - 1/2 matrices. The potential energy consists of a one-particle (two-level) part $\omega_0/2$ and two pair interaction terms $U(r)$ and $J(r)$, where U is a hard disc potential for particles with diameter d and $J(r) = J$ for $d < r < 1.5\,d$ and zero elsewhere. For motivation and literature background for these type of models see ref.[6]. In the adiabatic approximation (large M) we assume a separation of time scales for the translational degrees of freedom and the internal quantum states. An application of the Trotter formula results in the following expression for the partition fuction $Z_{N_1,V_1,T} = \lim_{P\to\infty} Z_P$ at temperature $T^* = (\beta J)^{-1}$, with

$$Z_P(\beta, N_1, V_1) = \frac{A_P^{N_1 P}}{\lambda^{2N_1} N_1!} \int d\mathbf{r}_1 \cdots \int d\mathbf{r}_{N_1} \exp[-\beta \sum_{i<j} U(r_{ij})]$$

$$\times \sum_{\{S\}} \exp[-\beta \sum_{i=1}^{N_1} \sum_{p=1}^{P} \left(K_P S_{i,p} S_{i,p+1} + \frac{1}{P} \sum_{i \neq j = 1}^{N_1} J(r_{ij}) S_{i,p} S_{j,p} \right)] \tag{4}$$

Here N_1 denotes the number of particles in one of the GEMC-boxes at volume V_1. In the GEMC simulation in addition all necessary volume and particle decompositions have to be taken into account properly for the full partition function [7]. The coefficients A_P and K_P are given by

$$A_P = [\frac{1}{2} sinh(\beta\omega_0/P)]^{1/2}, \quad K_P = \frac{1}{2\beta} ln[coth(\beta\omega_0/2P)] \,. \tag{5}$$

λ is the thermal de Broglie wavelength and the quantum chains have periodic boundary conditions w.r.t. P, and $S_{i,p} = \pm 1$.
In order to study the systematic ensemble and size dependencies of the gas-liquid coexistence densities, which recently got much attention in the context of GEMC simulations [8], we chose the same interaction parameters as in ref. [6]. The Gibbs ensemble simulations [7] were done with $N = 200$ particles and $P/\beta J = 40$. A typical run over 10^6 Monte Carlo steps (consisting of 200 attempted moves, 20 particle exchange and 1 volume change attempts) took about 14 CPU hours on a CRAY YMP.

Table 2: *Gas-liquid coexistence densities of the two-dimensional fluid with internal quantum states versus temperature. Comparison of Gibbs ensemble Monte Carlo results with those of an NVT- ensemble simulation and mean field predictions.*

T^*	Gibbs- MC		NVT- MC [6]		Mean Field	
	ρ_g^*	ρ_l^*	ρ_g^*	ρ_l^*	ρ_g^*	ρ_l^*
0.35	0.139	0.739	0.174	0.711	0.198	0.762
0.4	0.196	0.715	0.232	0.659	0.240	0.744
0.45	0.265	0.685	0.298	0.639	0.276	0.726
0.5	0.323	0.633	0.341	0.576	0.3	0.702
0.55	0.376	0.521	0.399	0.506	0.324	0.684

The results of the GEMC simulations for the gas-liquid coexistence densities are presented in table 2, together with the NVT- ensemble results of ref.[6] and mean field predictions. We note only a weak ensemble dependency of the results. The mean field theory predicts a too large coexistence region resulting in a 100% deviation from the GEMC results for the gas-liquid critical temperature T_c, which was found by GEMC to be about $T_c = 0.57$, in good agreement with NVT-ensemble results. A systematic study of the system size dependencies will be presented in a future work.

Acknowledgements

The authors thank K. Binder and D. Marx for useful discussions and the HLRZ/Jülich and the RHRZ Kaiserslautern for computer time. F.S. thanks the DFG for financial support (DFG Ni 259/6-1) and P.N. thanks the DFG for a Heisenberg fellowship.

References:

[1] T. Biben, J.P. Hansen, Phys. Rev. Lett. **66**, 2215 (1991); J. G. Amar, Mol.Phys. **67**, 739 (1989); D. Gazillo, G. Pastore, Chem. Phys. Lett. **159**, 388 (1989).

[2] A.Z. Panagiotopoulos, Mol. Phys. **61**, 813 (1987); Molec. Silmul. **9**, 1 (1992).

[3] K. Binder, Z. Phys. **B 43**, 119 (1981); M. Rovere, D.W. Heermann, K. Binder, Europhys. Lett. **6**, 585 (1988).

[4] M.-O. Ihm, Diplomarbeit, Mainz 1993.

[5] T.W. Melnyk, B.L. Sawford, Mol. Phys. 29, 891 (1975).

[6] D. Marx, P. Nielaba, K. Binder, Phys. Rev. Lett. 67, 3124 (1991); Phys. Rev. B 47, 7788 (1993).

[7] F. Schneider, Dissertation, Mainz.

[8] K.K. Mon, K. Binder, J. Chem. Phys. 96, 6999 (1992).

Optimization by Multicanonical Annealing and the Traveling-Salesman Problem

J. Lee[1] *and M.Y. Choi*[2]

[1]Supercomputer Computations Research Institute (SCRI), Florida State University, Tallahassee, FL 32306-4052, USA
[2]Department of Physics, Seoul National University, Seoul 151-742, Korea

Abstract. We demonstrate a powerful and general simulated annealing method to study combinatorial optimization problems. It combines the multicanonical method, which samples directly the microcanonical entropy of the system, with an elaborate but straightforward annealing scheme. The idea is to fully utilize the information about the local entropy obtained during short Monte Carlo simulations for optimization in an iterative fashion. We present results of an extensive investigation of the traveling salesman problem in a unit square.

1. Introduction

Many optimization problems that arise in various areas of science and engineering are hard to solve. One of the well-known classic examples is the traveling salesman problem (TSP), which consists of finding the shortest tour connecting N cities [1]. The TSP is widely believed to require computing time which grows exponentially with N. Perhaps the most impressive achievement in the TSP has been obtained by the branch-and-cut algorithm which has found *proven* optimal solutions in reasonable amounts of time for up to N=2392 [2]. In practice, however, near-optimal solutions are usually satisfactory, and it is more desirable to obtain such approximate solutions rather quickly.

A decade ago some common features between optimization problems and disordered spin systems were recognized, leading to the application of the simulated annealing (SA) method [3]. This method is based on Monte Carlo (MC) simulations. In reality, the SA may be outperformed by the best heuristic (deterministic) algorithms for a specific optimization problem. For example, it has been claimed that the iterative Lin-Kernighan (LK) algorithm finds better solutions of the TSP [4, 5]. Nevertheless, the SA is far superior to the heuristic algorithms in its *simplicity* and *versatility*. It is thus desirable to improve the SA so that it is *both efficient and versatile* by considering more efficient sampling algorithms and annealing schemes. Recently, significant progress has been achieved in the search for more efficient MC sampling algorithms [6, 7, 8, 9].

2. General Algorithm

The new optimization method [10], which we call *multicanonical annealing* is based on the multicanonical method (MM) [7]. In particular, we follow the

Springer Proceedings in Physics, Vol. 78
Computer Simulation Studies in Condensed-Matter Physics VII
Editors: D.P. Landau · K.K. Mon · H.-B. Schüttler

notation of Ref. [8]. The MM allows one to sample the density of states $\Omega(E)$ directly by imposing the appropriate detailed balance condition. The microcanonical entropy $S(E)$ is defined as $S(E)=\ln \Omega(E)$. [In the multicanonical notation, $S(E)=\beta(E)E-\alpha(E)$. $\alpha(E)$ and $\beta(E)$, which appear only here, should not be confused with α and β defined below.] The framework of the current optimization method is based on the feedback of information about the *local entropy* as described in the following procedure: (I) Assume the system is in a particular local minimum A. (II) Obtain new information on the local entropy S near A via a short MC sampling to update the old value of S. (III) Modify the transition probability of the MC move according to the new S so the system is driven out of A with a high probability and falls into a new local minimum A'. (IV) Iterate the three steps until a chosen convergence criterion is met.

In conventional SA, directed cooling is achieved by slowly lowering the temperature coupled to the function l to be minimized. Here, we place a hard wall potential on the other end of the annealing direction, rejecting all attempts beyond the wall. Annealing is achieved by moving the wall (in the annealing direction) along with the optimal l obtained thus far, in order to control the size of the sampling interval. We choose the size of the sampling interval to be fixed and large enough to allow important fluctuations throughout the annealing process, whereas the corresponding interval for conventional SA becomes smaller as the temperature is lowered (and eventually vanishes at zero temperature).

3. TSP in a Unit Square and Details

We consider the classic Euclidean problem of N cities randomly distributed inside a unit square for N up to 40000. The minimal tour length $l_{\rm m}(N)$ was shown [11] to scale as $l_{\rm m}(N)=\alpha\sqrt{N}$ in the large-N limit. However, there has not been much progress on the estimate of $\alpha(N) \equiv l_{\rm m}(N)/\sqrt{N}$ for a given N, not to mention the estimate of the limiting value $\alpha(N\rightarrow\infty)$. We describe the overall algorithm as follows:

(1) We start with an arbitrary tour configuration. In the case of the TSP, we have been able to use the configuration from 2-opt heuristics [12], which on average has about a 5% longer length than the final result. For conventional SA, the starting configuration is of longer length. This advantage is due to the fact that in the MM the ergodicity problem can be alleviated to some extent by helping the system escape quickly from local minima. We set the maximum allowed length $l_{\rm max}$ to be between one and ten percent longer than the result from 2-opt heuristics. Initially the entropy $S(l)$ is set to be a constant for all properly discretized lengths l. Observing that the distance between neighboring cities scales as $1/\sqrt{N}$, we multiply the tour length by $10\sqrt{N}$ and set it to an integer. We make a look-up table that lists the 20 nearest cities for each city.

(2) We use the 2-bond move, which reverses the sequence of cities for a chosen segment, in order to obtain trial tours. Suppose that the current tour is given by $(c_1, c_2, \cdots, c_N)$. For a randomly selected city c_i, we choose a city

c_j, one of the 20 nearest cities of c_i excluding the adjacent cities in the path (i.e., $j \neq i \pm 1$). We consider a 2-bond move which replaces the two bonds (c_i, c_{i+1}) and (c_j, c_{j+1}) by (c_i, c_j) and (c_{i+1}, c_{j+1}). We adopt the heat-bath MC algorithm by considering all 20 neighbors from the look-up table. The new tour is chosen out of these trial tours and the current tour with the probability $\exp(-S_i)/\sum_i \exp(-S_i)$. If the new length is larger than $l_{\max}$, the new tour is rejected (due to the hard wall potential).

(3) We repeat step (2) for M MC sweeps (MCS) while obtaining a histogram of the distribution of l from the MN events. Typically we have used M=25.

(4) The new estimate of $S(l)$ is now given by

$$S(l) = \begin{cases} S(l) & \text{for } H(l) = 0, \\ S(l) + \ln H(l) & \text{otherwise,} \end{cases} \tag{1}$$

where $H(l)$ is the (unnormalized) histogram of the sampling. If $l_{\min}$ is the minimum tour length obtained so far, then for $S(l)$ with $l < l_{\min}$, $S(l)$ is obtained by linear extrapolation with the slope $[S(l_{\max}) - S(l_{\min})]/(l_{\max} - l_{\min})$. This procedure expedites the movement of the system in the desired direction.

(5) The new $l_{\max}$ is chosen so that $l_{\min}$ and $l_{\max}$ form a new interval with a fixed difference of $5\sqrt{N}$. If the length l_c of the current tour is longer than the new $l_{\max}$, the latter is set equal to l_c.

This defines one iteration. With the new estimate of $S(l)$ and $l_{\max}$ we repeat the steps (2)–(5) until it fails to find a better tour for 20 consecutive iterations.

4. Results

We summarize the performance of the multicanonical annealing method in Table I. Note that only 22000 total MCS have been used for $N = 40000$. It is important to obtain information about the entropy of the current local minimum quickly, so that the system can easily escape from it. This is based on the fact that the systems under consideration are not ergodic, especially near ground states. For the case of the TSP, the tour length alone cannot serve as a relevant order parameter, that is, it is likely that there exist tours with comparable lengths that do not belong to the same local minimum. This necessitates very frequent updates of the information about the local entropy.

In Fig. 1 we plot $\alpha(N) \equiv l/\sqrt{N}$ as a function of $1/\sqrt{N}$, assuming $l(N) = \alpha\sqrt{N} + \beta$ [11], where α is $\alpha(N \to \infty)$. A commonly cited value of α is 0.749 [11]. Recently, Krauth and Mèzard used the cavity method to obtain α=0.7251 for a different version of the TSP, where the symmetric distance matrix is distributed uniformly between 0 and 1 [13]. A special case of that problem was then argued to be mapped onto the current problem. Our value for α=0.7211(3) agrees reasonably well with 0.7251 and rules out the value 0.749. The commonly cited value $\alpha(N)$=0.749 [1, 14, 15] is neither accurate nor appropriate since $\alpha(N)$ depends on N. To our knowledge, this is the first time that a systematic finite-size scaling analysis [16] for the TSP has been performed.

Table 1: Performance of the multicanonical annealing for the TSP with N cities in a unit square. n is the number of independent random configurations. σ is the standard deviation of the mean $\alpha(N)$.

N	n	$\alpha(N) = l/\sqrt{N}$	σ	MCS
50	1300	0.8075	0.0010	1100
64	900	0.7968	0.0010	1300
81	700	0.7886	0.0010	1400
100	600	0.7802	0.0009	1500
121	500	0.7754	0.0009	1700
144	400	0.7704	0.0010	1800
200	300	0.7639	0.0009	1900
256	200	0.7594	0.0010	2100
400	120	0.7515	0.0010	2500
900	80	0.7418	0.0007	3300
1600	50	0.7362	0.0007	4700
2500	25	0.7331	0.0008	5400
10000	8	0.7278	0.0007	8600
40000	4	0.7239	0.0004	22000

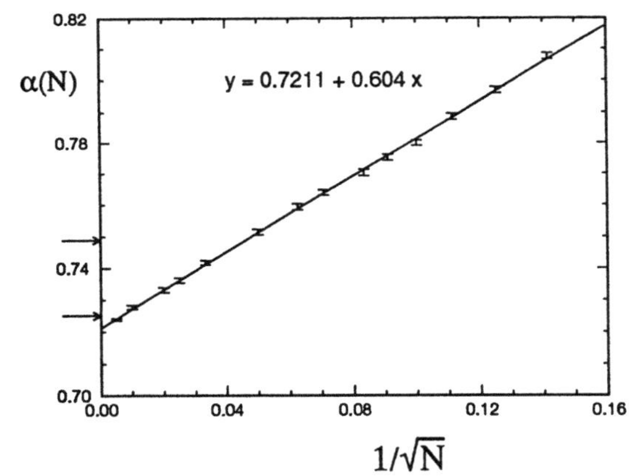

Figure 1: Arrows indicate the commonly cited value 0.749 in Ref. [11] and the predicted value 0.7251 in Ref. [13]. Even though 0.7251 is in fair agreement with our estimate 0.7211, the exact value should be somewhat smaller than 0.7211, since this work provides only an upper bound.

Although our data serve only as an upper bound for the TSP in a unit square, the results are overwhelmingly better than existing large scale TSP studies [14, 17] using the conventional SA method. It should also be noted that the direct approach by the conventional SA method takes too long to obtain reasonable results for the large scale TSP, and some tricks such as the divide-

and-conquer strategy are often used [1, 14]. The only source we have been able to find for comparison with our results is Ref. [14], where the tour length l=76.3 was obtained for N=10000 by dividing the data into 4096 subdivisions. Our average result l=72.8 is shorter than l=76.3 by almost the perimeter of the unit square. We also find that the computer time used is quite small compared to the conventional SA method [15, 17, 18]. For $N > 2500$, we had to compute the inter-city distances each time due to the limited memory. Notwithstanding this restriction, our results for N=10000 took about 7 CPU hours on a single ibm340h workstation, corresponding to 8600 MCS.

5. Discussion

Even though our results represent a major improvement over those obtained using the conventional SA approach, there are also other ingenious approaches based on heuristics. For the same realization of N=31623, with 20000 MCS we found a tour of $\alpha(N)$=0.723, which is only slightly better than the average result of the LK heuristics [4] by Johnson; the latter took about 67 seconds in a large-memory workstation. Using the iterative LK [5] procedure, Johnson has obtained a tour of $\alpha(N)$=0.716 with 16 hours of cpu time. Even so, we believe that our annealing method is very powerful, not only due to its versatility but also its efficiency. Our approach is certainly more general than a particular heuristic algorithm such as the iterative LK algorithm. It would be very interesting to apply our annealing method to various optimization problems with more complicated cost functions and to compare the results with those from problem-specific heuristics.

Acknowledgements. We are grateful to D. Johnson for informing us of his unpublished results, and to B. A. Berg for discussions about the multicanonical method. We also would like to thank B. Gorman, H. Meirovitch, M. A. Novotny, and P. A. Rikvold for useful discussions. JL was supported by the DOE through a grant to the SCRI, Florida State University (Contract DE-FC05-85ER25000). MYC was supported in part by the Ministry of Education of Korea, by the NSF (Grant DMR-9220733), and by the KOSEF through a grant to the CTP of Seoul National University.

References

[1] *The Traveling Salesman Problem: A Guided Tour of Combinatorial Optimization*, edited by E. L. Lawler, J. K. Lenstra, A. H. G. Ringnooy Kan, and D. B. Shmoys (Wiley, Chichester, 1985).

[2] M. Padberg and G. Rinaldi, SIAM Rev. **33**, 60 (1991).

[3] S. Kirkpatrick, C. D. Jr. Gelatt, and M. P. Vecchi, Science **220**, 671 (1983); V. Cerny, J. Optimization Theory Appl. **45**, 41 (1985).
[4] S. Lin and B. W. Kernighan, Oper. Res. **21**, 498 (1973).
[5] D. Johnson, Nature **330**, 525 (1987); *Proceedings of the 17th Colloquium on Automata, Language, and Programming* (Springer-Verlag, Berlin, 1990), pp. 446-461.
[6] J. P. Valleau and D. N. Card, J. Chem. Phys. **57**, 5457 (1972); G. M. Torrie and J. P. Valleau, J. Comput. Phys. **23**, 187 (1977).
[7] B. A. Berg and T. Neuhaus, Phys. Lett. B**267**, 249 (1991); Phys. Rev. Lett. **68**, 9 (1992); B. A. Berg and T. Celik, Phys. Rev. Lett. **69**, 2292 (1992); B. A. Berg, Int. J. Mod. Phys. C**3**, 1083 (1992).
[8] J. Lee, Phys. Rev. Lett. **71**, 211 (1993); **71**, 2353 (1993) (erratum).
[9] A. Hüller, Z. Phys. B **88**, 79 (1992).
[10] J. Lee and M. Y. Choi, preprints, FSU-SCRI-93-107.
[11] J. Beardwood, J. H. Halton, and J. M. Hammersley, Proc. Camb. Philo. Soc. **55**, 299 (1959); J. M. Steele, Ann. Prob. **9**, 365 (1981).
[12] G. A. Croes, Oper. Res. **6**, 791 (1958); S. Lin, Bell Syst. Tech. J. **44**, 2245 (1965).
[13] W. Krauth and M. Mèzard, Europhys. Lett. **8**, 213 (1989).
[14] E. Bonomi and J.-L. Lutton, SIAM Rev. **26**, 551 (1984).
[15] B. K. Ambati, J. Ambati, and M. M. Mokhtar,
[16] See, e.g., *Finite Size Scaling and Numerical Simulation of Statistical Systems*, edited by Privman, V. (World Scientific, Singapore, 1990).
[17] F. Favata and R. Walker, R., Biol. Cybern. **64**, 463 (1991).
[18] R. Durbin and D. Willshaw, Nature **336**, 689 (1987). Biol. Cybern. **65**, 31 (1991).

Numerical Studies of Metastability in Ising Systems: Short-Range Versus Long-Range Interactions

B.M. Gorman and C.C.A. Günther

Supercomputer Computations Research Institute, Department of Physics, and Center for Materials Research and Technology,
Florida State University, Tallahassee, FL 32306, USA

Abstract. We apply field-theoretical and transfer-matrix methods to study homogeneous nucleation in Ising ferromagnets both with short-range and with long-range interactions. A "constrained" free-energy density computed directly from the transfer matrix is compared with the analytically continued free-energy density from field-theoretical treatments and with Monte-Carlo data.

1. Metastability and Interaction Range

In a study of thermally activated nucleation, Langer [1] showed that the nucleation rate $\Gamma = \frac{\beta\kappa}{\pi}|\mathrm{Im}\widetilde{F}|$, where $\widetilde{F}$ is the free energy analytically continued around the condensation point, and κ contains all dependence on the dynamics. A correspondence between $\widetilde{F}$ and the analytic continuation of the dominant transfer-matrix eigenvalue has been supported by numerical studies of the 2D Ising ferromagnet [2, 3, 4] and of a quasi-one-dimensional Ising (Q1DI) ferromagnet with long-range interactions [5, 6]. Here we report recent applications [7, 8] to both models of the constrained-transfer-matrix (CTM) method [6], in which a "constrained" free-energy density f_α is associated with the metastable phase. We compare f_α to the analytically continued free-energy density $\tilde{f}$, and we compare and contrast the implied nucleation properties of models with short- and long-range interactions (SRI and LRI).

In the field-theoretical description of nucleation, a coarse-grained free-energy functional is defined in terms of a local order parameter ϕ, and nucleation is assumed to proceed through growth of noninteracting droplets of an energetically favored phase ϕ_+ in a sea of the metastable phase ϕ_- [9]. Although the noninteracting-droplet picture is usually valid for both SRI and LRI models, the droplet geometries differ. In the limit as the applied magnetic field $H \rightarrow 0$, the droplets in both models have separate length scales $R_c \gg \delta_S$ for their critical radius R_c and surface depth δ_S [9, 10]. However, R_c rapidly decreases as H departs from zero. Since δ_S is much larger for LRI models than for SRI models, R_c and δ_S become a single length scale at a much smaller H. Another effect of the larger δ_S is that a critical droplet in an LRI model nucleates by increasing the concentration of ϕ_+ at its center [10, 11], whereas in an SRI model it nucleates by adding mass to its surface [9, 12].

The 2D Ising model, an example of an SRI model, is defined by the Hamiltonian $\mathcal{H} = -J\Sigma_{\langle i,j\rangle}\sigma_i\sigma_j - H\Sigma_i\sigma_i$, where $\sigma_i = \pm 1$ is the spin at site i, and the

Springer Proceedings in Physics, Vol. 78
Computer Simulation Studies in Condensed-Matter Physics VII
Editors: D.P. Landau · K.K. Mon · H.-B. Schüttler

first sum is over nearest-neighbor pairs on the square lattice. In this case $\tilde{f}$ is best approximated in the limit $H \to 0$, giving [9, 12]

$$|\mathrm{Im}\tilde{f}| = B(T)|H|^{b} \exp\left(-\beta\Xi(T)|H|^{-1}\left[1+O(H^2)\right]\right) . \tag{1}$$

where $\Xi(T)$ is related to the exact zero-field surface tension and the equilibrium spontaneous magnetization. The universal exponent b=1 if Goldstone excitations on the surface of the critical droplet are present.

The Q1DI model [5], an example of an LRI model, is defined as a one-dimensional chain of L layers, each containing N Ising spins interacting ferromagnetically with the mean-field of each neighboring layer, so that the interaction range $R_{\mathrm{I}} \sim N$. The Hamiltonian is $\mathcal{H} = -N\Sigma_i(J\phi_i\phi_{i+1} + H\phi_i)$, where ϕ_i is the magnetization density at layer i. This model exhibits a sharp spinodal H_{s}, and $\tilde{f}$ is best approximated in the limit $\lambda = |H| - |H_{\mathrm{s}}| \to 0^-$, giving [8]

$$|\mathrm{Im}\tilde{f}| = A(T)N^{-1/2}|\lambda|^{7/8} \exp\left(-\beta\Phi(T)N|\lambda|^{5/4}\left[1+O(\lambda^{1/2})\right]\right) . \tag{2}$$

2. Numerical Results

The CTM method [6] extends the usual concept of the transfer matrix to constrained nonequilibrium states. For each eigenvalue λ_α of $\mathbf{T}$ we define a "constrained" transfer matrix (CTM) $\mathbf{T}_\alpha$ so that "constrained" probability densities are defined in analogy with the equilibrium (α=0) case:

$$P_\alpha(x_i, x_{i+k}) \quad = \quad \langle\alpha|x_i\rangle\langle x_i|(\lambda_\alpha^{-1}\mathbf{T}_\alpha)^{|k|}|x_{i+k}\rangle\langle x_{i+k}|\alpha\rangle . \tag{3}$$

We choose $\mathbf{T}_\alpha$ to commute with $\mathbf{T}$, with each eigenvalue $\lambda > \lambda_\alpha$ reweighted to λ_α^2/λ, in order to ensure that the constrained system is characterized by P_α. A "constrained" free-energy density $f_\alpha = \langle\mathcal{H}\rangle_\alpha - TS_\alpha$ is defined by replacing equlibrium probability weights in $\langle\mathcal{H}\rangle$ and S with P_α:

$$f_\alpha = -\frac{\ln|\lambda_\alpha|}{\beta N} + \frac{1}{\beta N}\sum_{x_i, x_{i+1}} P_\alpha(x_i, x_{i+1}) \mathrm{Ln}\left(\frac{\langle x_i|\mathbf{T}_\alpha|x_{i+1}\rangle}{\langle x_i|\mathbf{T}|x_{i+1}\rangle}\right) , \tag{4}$$

where $\mathrm{Ln}(z)$ is the principal branch of $\ln(z)$. The second term here is complex-valued in general, since $\mathbf{T}_\alpha$ is not in general a positive matrix.

For a unique value of α at a given T and H, $\mathrm{Re}f_\alpha$ very closely approximates $\mathrm{Re}\tilde{f}$, and $\mathrm{Im}f_\alpha$ is very close to zero. Due to strong eigenvector mixing, $\mathrm{Im}f_\alpha$ exhibits a lobe structure. We select values of H for which the effects of eigenvector mixing are minimal, and we identify the envelope connecting these points with $\mathrm{Im}f_{\mathrm{ms}}$, the imaginary part of the constrained free-energy density for the metastable state.

2.1 The 2D Ising Model

Figure 1(a) shows $|\mathrm{Im} f_\alpha|$ vs. J/H for the "metastable" α. In weak fields, $R_c > N$, so the free-energy cost ΔF of creating a critical droplet is simply that of an interface dividing the system, and thus $\Delta F \sim N$. Consequently, $\mathrm{Im} f_{\mathrm{ms}}$ decays exponentially with N, although the range of fields over which this scaling holds vanishes as N^{-1}. For $2/(N-2) \leq H/J \leq 2$, $\mathrm{Im} f_{\mathrm{ms}}$ is independent of N and decays exponentially with $|H|^{-1}$, in agreement with Eq. (1). In Fig. 1(a), this is demonstrated by the straight line, which was drawn to connect the two minima marked by (•) for $N=10$. Its slope gives a rough estimate for $\Xi(T)$ in Eq. (1). At H_{MFSP}, the metastable single-phase correlation length equals the size of the critical droplet, so we expect droplet theory to break down [7]. Indeed, the minima start to deviate strongly from the straight line around H_{MFSP}.

In order to determine $\Xi(T)$ and b from our CTM results, we performed least-squares fits of $\ln|\mathrm{Im} f_{\mathrm{ms}}|$ to the form of Eq. (1) [7]. Figure 1(b) shows the excellent agreement between the equilibrium value of $\Xi(T)$ (solid line) and its value obtained from a two-parameter fit (∘) with b fixed to 1. A more thorough analysis of the results of our least-squares fits gives estimates of b consistent with the presence of Goldstone modes on the critical droplet surface [12]. We also compared $\Xi(T)$ to direct lifetime measurements from Monte-Carlo simulations (× [13] and $*$ [14] in Fig. 1(b)) using Langer's formula and Eq. (1). For $T/J \geq 0.7$, the Monte-Carlo results are in excellent agreement with those obtained from the exact zero-field surface tension and the CTM method. For low T, the size of the critical droplets is small at all H that can be resolved with the Monte-Carlo method. The discreteness of the lattice then causes the estimates for $\Xi(T)$ to be consistently too large [13].

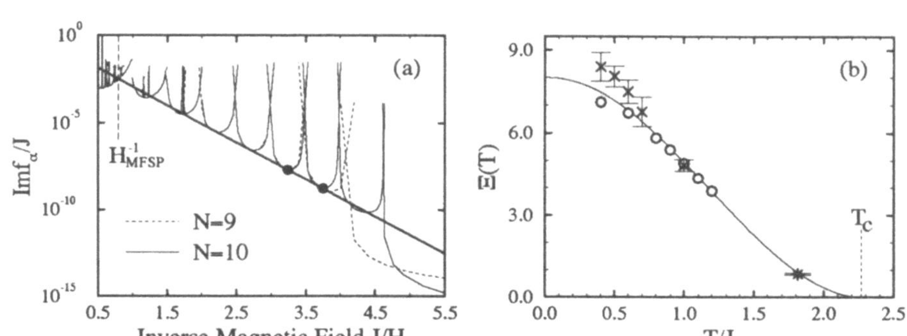

Figure 1: (a) $|\mathrm{Im} f_\alpha|$ for $T/J=1$ ($T/T_c \approx 0.441$), and (b) $\Xi(T)$ for the 2D Ising ferromagnet.

2.2 The Q1DI Model

Figure 2(a) shows $|\text{Im} f_\alpha|$ vs. $H/2J$ for the Q1DI model. The values of H selected to determine $\text{Im} f_{\text{ms}}$ are marked by the (+) symbols. By inspecting the envelopes for many system sizes, we observed roughly an exponential dependence on N well inside the spinodal and a crossover to power-law scaling near H_s. Assuming a scaling form given by Eq. (2), we estimated the free energy cost of forming the critical droplet $\Delta F \approx \beta^{-1} \ln |\text{Im} f_{\text{ms},N}(T,H)|$. Extrapolated finite-$N$ estimates for $\Delta F/N$ are shown in Fig. 2(b). The agreement between the extrapolated CTM estimates and the field-theoretical results strongly supports the consistency between the CTM method and the droplet theory of nucleation.

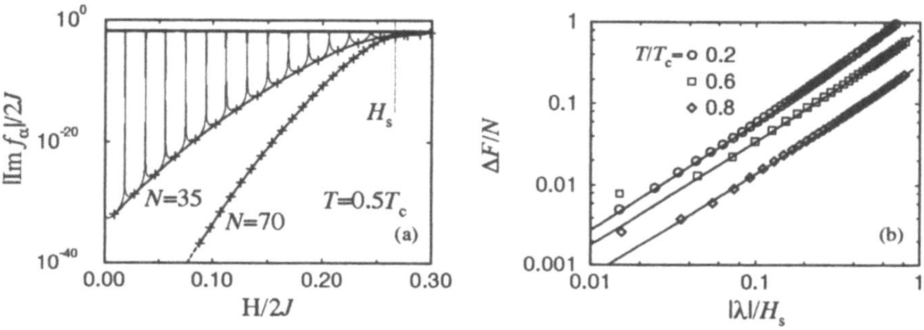

Figure 2: (a) Branches of $\ln|\text{Im} f_\alpha|$ for the Q1DI model. Also shown are envelopes for two sizes. (b) Extrapolated finite-N estimates for $\Delta F/N$. The lines indicate the field-theoretical result.

3. Conclusion

The constrained-transfer-matrix (CTM) method [6] applied to the 2D Ising and Q1DI ferromagnets illustrates the differences between models of different interaction ranges. In both cases, the imaginary part of the constrained free-energy density calculated from the CTM showed strong agreement with the analytically continued free-energy density from field-theoretical droplet models. Both the dependence on interaction range in the Q1DI model and the dependence on the surface free energy and critical droplet shape in the 2D Ising model were captured by the CTM method. Clearly this method shows promise in characterizing metastable phases. As a nonperturbative method, it treats all possible fluctuations in a single calculation, and hence, unlike field-theoretical approaches, does not require any particular assumptions about the detailed structure of the free-energy functional.

Acknowledgements. We thank P. A. Rikvold, M. A. Novotny, and T. Fiig for useful conversations, and the authors of Refs. [13, 14] for use of their data prior to publication. This work was supported in part by FSU through the Supercomputer Computations Research Institute (US DOE Contract DE-FC05-85ER25000), through the Center for Materials Research and Technology, and through Cray Y-MP supercomputer time, and by NSF Grant DMR-9013107.

References

[1] J. S. Langer, Ann. Phys. **54**, 258 (1969).
[2] R. J. McCraw and L. S. Schulman, J. Stat. Phys. **18**, 293 (1978).
[3] V. Privman and L. S. Schulman, J. Phys. A **15**, L231 (1982).
[4] V. Privman and L. S. Schulman, J. Stat. Phys. **31**, 205 (1982).
[5] M. A. Novotny, W. Klein, and P. A. Rikvold, Phys. Rev. B **33**, 7729 (1986).
[6] P. A. Rikvold, Prog. Theor. Phys. Suppl. **99**, 95 (1989).
[7] C. C. A. Günther, P. A. Rikvold, and M. A. Novotny, Phys. Rev. Lett. **71**, 3898 (1993); work in progress.
[8] B. M. Gorman, P. A. Rikvold, and M. A. Novotny, Phys. Rev. E, accepted.
[9] J. S. Langer, Ann. Phys. **41**, 108 (1967).
[10] M. Büttiker and R. Landauer, Phys. Rev. A **23**, 1397 (1981).
[11] C. Unger and W. Klein, Phys. Rev. B **29**, 2698 (1984).
[12] N. J. Günther, D. A. Nicole, and D. J. Wallace, J. Phys. A **13**, 1755 (1980).
[13] M. A. Novotny, in these proceedings.
[14] P. A. Rikvold, H. Tomita, S. Miyashita, and S. W. Sides, submitted to Phys. Rev. E.

Study of Frustrated Systems Via Dissipative Maps

G.S. Canright[1,2] *and G.I. Watson*[1,2,*]

[1]Department of Physics and Astronomy, The University of Tennessee, Knoxville, TN 37996-1200, USA
[2]Solid State Division, Oak Ridge National Labortory, Oak Ridge, TN 37831, USA
*Present address: Rutherford Appleton Laboratory, Chilton, Didcot, Oxon OX11 0QX, UK

Abstract. We briefly describe a method for finding complex metastable states for frustrated classical systems, using a dissipative dynamical map. We show by example that this technique can give states which are very competitive in energy with the ground state, and with other, simpler states which are found by exhaustive search.

1. Introduction: frustration

We wish to consider "frustrated" problems in classical equilibrium statistical mechanics. Suppose we have an energy function

$$F = \sum_{\text{pairs } ij} f_{|i-j|}(\theta_i, \theta_j) \ , \qquad (1)$$

where the set $\{\theta_i\}$ represents the degrees of freedom of the problem. We call this function (and the system it represents) *frustrated* if it is impossible to choose a set of coordinates $\{\theta_i\}$ such that every term in F is simultaneously minimized.

Since the equilibrium problem is precisely to minimize F, then clearly when F is frustrated the equilibrium problem is in general nontrivial. Let us now further restrict our consideration to problems in which the frustration arises without any disorder in F. Even in this restricted class of problems, it is in general very difficult to deduce the absolute minimum of F. Furthermore, as F becomes increasingly frustrated, the importance of the absolute minimum ("ground state") diminishes relative to that of other local minima (metastable states), due to the tendency for large activation barriers to appear between minima (F becomes "crumpled"). Hence we broaden the problem, in the case of frustration, to one of finding a representative set of low-energy ("competitive") minima—a set which includes but is not limited to the ground state.

One interesting technique which has been applied to such problems involves writing the minimization problem as a mapping [1]. To pursue this idea we

Springer Proceedings in Physics, Vol. 78
Computer Simulation Studies in Condensed-Matter Physics VII
Editors: D.P. Landau · K.K. Mon · H.-B. Schüttler

restrict ourselves further to one-dimensional problems. Then the minimization problem for (1) may be written as an infinite set of coupled equations

$$\frac{\partial}{\partial \theta_i} F(\ldots \theta_{i-2}, \theta_{i-1}, \theta_i, \theta_{i+1}, \theta_{i+2}, \ldots) = 0 \ . \tag{2}$$

Here the coordinates are listed in 1D order. Assuming the interaction in F extends to the rth neighbor, there are $2r+1$ coordinates in each of (2). The ith equation may then be rewritten as an equation for θ_{i+r} in terms of the $2r$ previous θ's. In this light, knowledge of those previous $2r$ values for θ will determine θ_{i+r}, from (2). Thus we obtain the $2r$-dimensional map

$$\begin{pmatrix} \theta_{i+r-1} \\ \theta_{i+r-2} \\ \vdots \\ \theta_{i-r} \end{pmatrix} \Longrightarrow \begin{pmatrix} \theta_{i+r} \\ \theta_{i+r-1} \\ \vdots \\ \theta_{i-r+1} \end{pmatrix} \ . \tag{3}$$

Trajectories of (3) then correspond to extrema of F, for whatever $2r$ boundary conditions are used to start the iteration. Such maps have two important features [1]: (i) they are volume-preserving; (ii) *stable* trajectories of the mapping correspond to *unstable* extrema of F, and vice versa. This latter feature makes it difficult to study complex minima of F [say, those represented by long-period or chaotic trajectories of (3)], although simpler structures given by fixed points of (3) can be located accurately.

2. Dissipative maps

We could solve problem (ii) above if we could replace the volume-preserving mapping (3) with a *dissipative* map—one whose orbits converge to *attractors* in the phase space—and if we could arrange that these attractors be low-energy minima (if not all the minima) of F. Such a mapping could then act as a 'magic bullet' which would efficiently locate those minima we want to study. This would include minima of *arbitrary* complexity. For example, with such a 'magic bullet' we could find *disordered* minima of F from chaotic attractors of the dissipative map, evaluate their energetics accurately, and otherwise characterize them in terms of correlations, etc.

We have discovered an approximate magic bullet, and applied it to a frustrated 1D problem of some physical interest [2]. In this problem F is of the form

$$F = \sum_{n=1}^{\infty} \sum_{n'=n+1}^{n+r} f_{|n-n'|}(|\theta_n - \theta_{n'}|) \quad , \tag{4}$$

where r is in principle infinite, but effectively finite (see Section 3). For this case our method may be described as follows. We believe [2] that there are regular solutions to (2) [with F as in (4)] of the form $\theta_j = j\phi$. We then consider a somewhat weaker condition: let $\overline{F}(\phi) \equiv F|_{\theta_j = j\phi}$, and assume that

$$\frac{\partial \overline{F}(\phi)}{\partial \phi} = 0 \quad , \tag{5}$$

that is, that there are uniform configurations which are stable against variation of the lattice constant. This is equivalent to

$$\sum_{\ell=1}^{r} \ell f_\ell'(\ell\phi) = 0 \quad , \tag{6}$$

where $f_n'(x) \equiv \partial f_n(x)/\partial x$.

Now we want a dissipative map which agrees with (6). We imagine (as with the volume-preserving map) "growing" the lattice from the left, but letting the choice θ_j^* of each new θ_j be a one-dimensional minimization problem:

$$G(\theta_j^*) = \min_{\theta_j} G(\theta_j) \quad \text{with} \quad G(\theta_j) = \sum_{n=j-1}^{n=j-r} g_{|j-n|}(|\theta_j - \theta_n|) \quad . \tag{7}$$

Since θ_j is then determined by the r previous values, this gives an r-dimensional map similar to (3). However, we choose a *fictitious* two-body potential g to replace the true potential f. It is then easily shown [2] that fixed points of the resulting map will correspond exactly to solutions to (6) (ie, to a uniform configuration stable with respect to variation in the uniform spacing ϕ) if we set

$$g_\ell(x) = \ell f_\ell(x) \quad . \tag{8}$$

In other words, we use knowledge about simple minima of F (eg, that uniform minima exist) to determine the fictitious potential G. The latter then gives rise to a dissipative map (7). Some of the attractors of this map will be (exactly) the 'simple' minima of F. However, we *suppose* that the mapping built from G will have other attractors of more complex structure, and that these other attractors will be *close to low-energy minima* of F. (The nearby exact minima may then be found by standard, purely local relaxation schemes.) This supposition has already been well supported by [2]; below we offer a further test of the idea.

3. Results: complex structures at low energy

In this section we wish to compare metastable states, found using the dissipative map (DM) described above, with other (small-period) metastable states found by an "exhaustive search" (ES) procedure. We will see (Fig. 1) that the DM can guide us to states of high complexity (including disordered states) which (a) could never be found by any exhaustive search procedure, and yet (b) are highly competitive in energy with those states which we can find by exhaustive search.

The problem we consider is that of Ref. [2]. Our two-body interaction f (representing repulsion between uniform layers of flux lines in a layered superconductor) is then $f_n(\alpha) = (\text{const}) \log(1 + e^{-4n\pi} - 2e^{-2n\pi} \cos 2\pi\alpha)$. Here

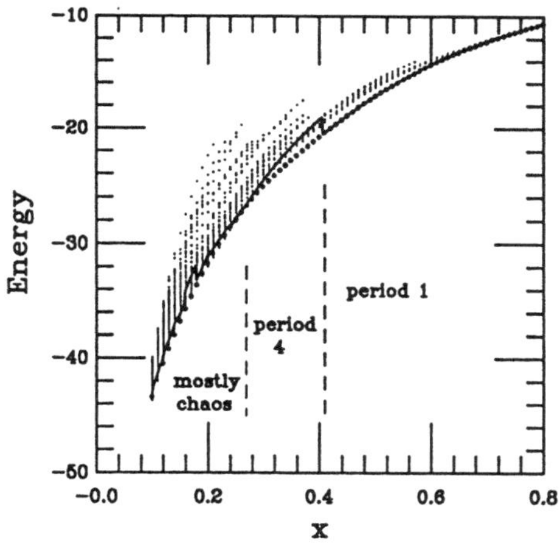

Figure 1. Energies for metastable states of a frustrated "spin chain", as a function of x ($\approx$ the inverse range of the antiferromagnetic interaction). *Small dots:* periodic minima found by "exhaustive search" (see text), for periods ≤ 8. *Large dots:* typical period-1 minima. *Solid line:* minima found using a dissipative map (followed by local relaxation). Structures of these minima are labelled in the figure. Note that, using the DM, we have found structures of great complexity (period $\rightarrow \infty$) which are nevertheless very low in energy.

$\alpha \equiv |\theta_i - \theta_j|$, $n = |i - j|$, and $0 \leq \theta_i < 1$. The parameter x is derived from the average flux line density. Small x makes f more long-ranged, and so (since f is uniformly repulsive) makes F more frustrated. For simplicity we may think of our 1D chain of layers as a classical XY spin chain, with the spin couplings long-ranged (range $\propto 1/x$) and uniformly antiferromagnetic.

Our ES procedure is similar to one described in Peyrard and Aubry [3]. To consider metastable states of period p, we study chains of length $K > 1/x$ (with K a multiple of p) and periodic boundary conditions. The configuration space for K spins can then be divided into sectors defined by the relative ordering of the spins in the unit interval. It is plausible that, for K larger than the range of interaction, each such sector contains (at most) a single minimum of F. We then seek that minimum by choosing a starting point in each sector (each spin getting one of the values ℓ/K, $\ell = 0, \ldots, K-1$) and relaxing (using standard routines) to the nearest local minimum.

A comparison of the ES energies with those found using the DM is given in Fig. 1. We see that, at small x, the DM allows us to find *disordered* (chaotic) minima of the energy (as well as large-period structures) [2] which are highly competitive with the simpler structures found by exhaustive search. Our method of finding these structures, using the DM, is *unique* (as far as we know) in this regard: that is, we know of no other method for finding these structures which is efficient enough to be practicable.

We are confident that this approach may be usefully applied to other frustrated 1D problems [4]. An extension to 2D (or higher) is of great interest. This may be possible in a restricted sense by using periodic boundary conditions in the other dimension(s); however this kills the possibility of observing structures which have an arbitrary period in either dimension. A more interesting problem (which we do not yet know how to solve) is to write an iterated mapping to represent the energetics for $D > 1$, if open BC are used in all dimensions.

Acknowledgments

We would like to thank David Phillippi for valuable contributions in the early stages of this research, and Ofer Biham for suggesting the exhaustive search. We also thank Ofer Biham, Ron Fox, and Andy Zangwill for helpful discussions and references. This work was supported by the University of Tennessee and the Division of Materials Science, U.S. Department of Energy under Contract No. DE-AC05-84OR21400 with Martin Marietta Energy Systems. GSC was supported in part by the NSF under grant # DMR-9101542.

References

1. See, for example, Per Bak, *Phys. Rev. Lett.* **46**, 791 (1981); Rahul Pandit and Michael Wortis, *Phys. Rev. B* **25**, 3226 (1982); and Ref. [3].

2. G.I. Watson and G.S. Canright, *Phys. Rev. B* **48**, 15950 (1993).

3. M. Peyrard and S. Aubry, *J. Phys. C* **16**, 1593 (1983).

4. We have explored limited applications of the method to the ANNNI problem [Julia Yeomans, *Solid State Physics* **41**, 151 (1988)] and to the long-range oscillatory potentials of metallic close-packed polytypes [A. Blandin, J. Friedel, and G. Saada, *J. Phys. (Paris) Colloq.* **27**, C3-128 (1966)].

Representation of Relaxation Phenomena by Lattice-Gas Models

P. Lallemand

Laboratoire de Physique Statistique, Ecole Normale Supérieure, F-75231 Paris Cedex 05, France

Abstract. After a brief introduction to lattice gases, the power spectral density of density fluctuations is computed in the general case. It is then shown that conditions can be found to represent acoustic relaxation. Next, transverse excitations are studied and it is shown that it is possible to obtain either purely relaxing or propagating shear waves. Taking intermediate conditions leads to viscoelastic behaviour.

1. Introduction

In an attempt to simplify molecular dynamics computations, Pomeau et al. in the 70's and later Frisch, Hasslacher and Pomeau[1] in the 80's introduced a model of fluids (later referred to as FHP) in which a collection of point particles move in a synchronous way along the links of a lattice of suitable symmetry. Thus instead of continuous motions, the lattice gas particles hop at each time interval from one vertex, or site, to another site of the lattice so that their velocity belongs to a finite set of dimension b, $\{\mathbf{c}_i,\ i = 1, ..., b\}$. This corresponds to the *propagation step* of the system. When particles meet, they redistribute their velocities during the *collision step*. A further simplification is brought by the fact that there is either 0 or 1 particle at each site with a given velocity.

The dynamics of the system is completely defined by the set of velocities and the rules (collision rules) during redistribution. These rules can either be deterministic or probabilistic and are such that some quantities are unchanged, like mass, linear momentum, energy or possibly other quantities.

2. Dynamics and correlation functions

Let us call $n_i(t, \mathbf{r}_l)$ the occupation number (0 or 1) of velocity $\mathbf{c}_i$ at site labelled l. The dynamical equation for n_i is given by

$$n_i(t+1, \mathbf{r}_l + \mathbf{c}_i) = n_i(t, \mathbf{r}_l) + \Delta_i \quad , \tag{1}$$

where Δ_i represents the change in the occupation of velocity i due to collisions.

For a statistical analysis of such a system, one assumes that the various velocities are uncorrelated so that the probability to have a given configuration

Springer Proceedings in Physics, Vol. 78
Computer Simulation Studies in Condensed-Matter Physics VII
Editors: D.P. Landau · K.K. Mon · H.-B. Schüttler

is the product of one particle functions. If we call f_i the mean occupation number, the dynamics of the system is given by a master equation

$$f_i(t+1, \mathbf{r}_l + \mathbf{c}_i) = f_i(t, \mathbf{r}_l) + \Omega_i \tag{2}$$

$$\text{with } \Omega_i = \sum_{s,s'} (s'_i - s_i) \mathcal{A}(s \to s') \prod_j^b f_j^{s_j} (1 - f_j)^{1-s_j} \quad , \tag{3}$$

where $\mathcal{A}(s \to s')$ is the probability that a collision takes an input state $s = \{s_1, ..., s_b\}$ over to the output state s'.

If we are interested in quantities like the dynamical structure factor, we consider the time evolution of an initial small perturbation of wavevector $\mathbf{k}$. To get results valid for long wavelength, we first linearize Eqn. 2 which involves the linearized collision operator $\mathcal{M}_{ij}$ that can be derived from Eqn. 3. The eigenvectors and eigenvalues of $\mathcal{M}$ play a particular role. Eigenvectors with zero eigenvalues correspond to hydrodynamic modes (or soft modes). Non-zero eigenvalues are related to transport coefficients or are irrelevant in the small k limit.

A priori, quantities that are conserved in collisions correspond to zero eigenvalues. In continuous systems none other appear. This may not be the case for lattice gases where unphysical quantities are sometimes conserved : they are called "spurious modes" following the work of Zanetti et al.[2].

3. Acoustic relaxation

We shall limit the analysis to 2-D. In such a case, the original FHP model with $b = 6$ behaves like a gas with acoustic excitations of frequency $\omega = c_s k$, where $c_s = \sqrt{1/2}$ is the velocity of sound. If we add "rest" particles of mass 1 so that $b = 7$, the velocity of sound becomes $c'_s = \sqrt{3/7}$. We can thus try and build a model where the speed of sound waves varies continuously from c'_s to c_s. For this purpose we consider the following model[3]. We take FHP with rest particles and use for the collision rules a mixture of the 6 and 7 bits models.

$$\mathcal{A}(s \to s') = p\mathcal{A}_7(s \to s') + (1-p)\mathcal{A}_6(s \to s') \quad , \tag{4}$$

where p is the probability of having a collision of type 7 and $1-p$ the probability of having a collision of type 6.

Detailed computations lead to a dispersion equation of sound modes that is identical to that considered in the Maxwell model of a fluid with a relaxation time τ for which a frequency dependent bulk viscosity can be used[4].

$$\zeta(\omega) = \frac{(c_\infty^2 - c_0^2)\tau}{1 + i\omega\tau} \quad \text{with, in our case, } c_\infty^2 = 1/2 \text{ and } c_0^2 = 3/7.$$

Numerical experiments[3] on the lattice gas discussed here show the same features as in real fluids with a single relaxation process. They exhibit acoustic dispersion for longitudinal excitations of frequency of the order of $1/\tau$ and their dynamical structure factor includes a central line of width $\simeq 1/\tau$, known as the Mountain line[5].

4. Viscoelasticity

The model discussed above shows no particular behavior of the shear viscosity. We thus have first to find means to influence the relaxation of transverse excitations. A possible way was presented in ref 6 where additional variables were added to the usual FHP model in such a way that new soft modes appear.

Here we consider a 2-D model with the usual 6 velocities at 60° from one another and include at each site an index K running from 1 to 6. This new variable is attached to sites of the lattice. It can only evolve during the collision step. The rules for collisions are such that the following relationships are satisfied

$$\sum_i n_i = \sum_i n_i', \ \sum_i \mathbf{c}_i n_i = \sum_i \mathbf{c}_i n_i', \ \sum_i c_i^2 n_i = \sum_i c_i^2 n_i', \tag{7}$$

$$\sum_i Q^i_{xx} n_i + F(K) = \sum_i Q^i_{xx} n_i' + F(K'), \tag{8}$$

$$\sum_i Q^i_{xy} n_i + G(K) = \sum_i Q^i_{xy} n_i' + G(K'), \tag{9}$$

where Q_{xx} and Q_{xy} are components of the stress tensor and F and G are components of a second rank tensor that depend upon K. The two new relations represent conservation of a second rank tensor and couple translation and molecular alignment as can be thought of a liquid made of elongated molecules. This coupling leads to transverse modes that can propagate with a phase velocity $c_\perp$.[6]

Other ways to obtain shear waves can be found from variants of the preceding model. One way is the following : take the hexagonal lattice of the FHP model and consider particles with speed 1 (motion to the nearest neighbour) and $\sqrt{3}$ (motion to the second-nearest neighbour). Collisions satisfy the conservation of mass, momentum and energy(as in Eqn. 5 to 7) together with that of the stress tensor, so that Eqn. 8 and 9 are replaced by

$$\sum_{i=1}^{12}(c_{ix}c_{ix} - \frac{1}{2}c_i^2)n_i = \sum_{i=1}^{12}(c_{ix}c_{ix} - \frac{1}{2}c_i^2)n_i' \quad , \tag{10}$$

$$\sum_{i=1}^{12} c_{ix}c_{iy}n_i = \sum_{i=1}^{12} c_{ix}c_{iy}n_i' \quad . \tag{11}$$

If we analyze the 2^{12} possible states, we find that 3408 can change in collisions which only conserve mass, momentum and energy, whereas only 612 can change if in addition Eqn. 10 and 11 are satisfied.

As in section 3, we shall consider two models : $\mathcal{M}_F$ with the full set of 3408 collisions and $\mathcal{M}_R$ with the reduced set of 612 cases. The "full" model behaves like the FHP model, except that there is in addition a diffusive mode corresponding to the fact that are two independent populations. (Note that to represent thermal phenomena additional velocities are needed to allow exchange between different speeds[7]). The reduced model exhibits propagating shear modes.

We now introduce a composite model $\mathcal{M} = (1-p)\mathcal{M}_R + p\mathcal{M}_F$. A detailed analysis of this new model shows that the shear viscosity is a function of the frequency and that depending upon values of k and p, transverse excitations are either under- or overdamped as expected in a real fluid whose shear viscosity exhibits relaxation. Obtaining results that are independent of the orientation of $\mathbf{k}$ is a little tricky, but can be achieved for a suitable choice of the ratio of the populations of velocities 1 and $\sqrt{3}$.

5. Conclusion

In this short account we have quoted results concerning the possibility of using lattice gases to simulate 2-D fluid flows with either a shear or a bulk viscosity that depends upon the frequency as in viscoelastic fluids. Extensions to 3-D is straightforward for bulk viscosity using the FCHC lattice gas[8] with "centers". It is found however that the ideas developped in section 4 cannot be used for the 24-bit FCHC lattice gas due to the appearance of spurious modes leading to undamped transverse waves for particular directions of the wavector. The corresponding anisotropy in the attenuation of shear modes can be reduced by taking a simple cubic model with velocities : 0, 1, $\sqrt{2}$ and $\sqrt{3}$ along the ideas of section 4.

References

1 - U. Frisch, B. Hasslacher and Y. Pomeau, Phys. Rev. Lett., **56**, 1505, (1986).

2 - G. Zanetti, Phys. Rev., **A 40**, 1539, (1989).

3 - P. Grosfils and P. Lallemand, Europhys. Lett., **24**, 473, (1993).

4 - K.F. Herzfeld and T. Litovitz, *Absorption and Dispersion of Ultrasonic Waves*, Academic Press, New York, (1959).

5 - R.D. Mountain, J. Res. Natl. Bur. Stands., **70A**, 207, (1966).

6 - D. d'Humières and P. Lallemand, Comp. Rend. Acad. Sci. Paris, **II 317**, 997, (1993).

7 - P. Grosfils, J.P. Boon and P. Lallemand, Phys. Rev. Lett., **68**, 1077, (1992).

8 - D. d'Humières, P. Lallemand and U. Frisch, Europhys. Lett., **2**, 291, (1986).

Some Monte-Carlo Studies of the Ising Model

G.A. Baker, Jr. and J.J. Erpenbeck

Theoretical Division, Los Alamos National Laboratory,
University of California, Los Alamos, NM 87545, USA

Abstract The Markov property method of speeding up standard Monte Carlo methods can yield about an order of magnitude improvement for the two-dimensional Ising model. We also study the effects of finite size in the three-dimensional Ising model on the computation of the renormalized coupling constant and find, so long as the correlation length is less that about one-tenth the system width, that an accuracy of one percent or better can be expected.

1. Introduction

A critical-phenomena problem which remains unresolved is the proper analysis of the behavior of the renormalized coupling constant in the three-dimensional Ising model, and the attendant problems of the anomalous dimension of the vacuum, hyperscaling and universality. In this paper we report two studies which are intended to form a basis for an adequate numerical approach to the resolution of this problem. The first is the investigation of a Markov property method intended to speed up Monte Carlo simulations. It was introduced and tried on the one-dimensional Ising model by Baker [1]. Here we examine its effectiveness on the two-dimensional Ising model. We find that it can provide about an order of magnitude speedup in this case. Secondly we investigate in the three-dimensional Ising model the approach of the finite-system, renormalized coupling-constant, $g_r(K;L)$, to the infinite-system limit. We calculate values of $g_r(K;L)$ and the correlation length $\xi_L(K)$ for systems of width $L \leq 100$ by either Monte Carlo or exact summation. We find $(K/K_c)^{3/2}g_r(K;L)$ as a function of $\xi_L(K)/L$ to be approximately independent of L and that, for $\xi_L(K) < L/10$, $g_r(K;L)$ should agree with $g_r(K;\infty)$ to within at least one-percent.

2. Auto-Correlation Times in Two Dimensions.

In some previous work the idea was introduced [1] of using the Markov property to speed up the Monte Carlo evaluation of thermodynamic functions in one dimension and to do [2] exact calculations on two-dimensional, finite-sized Ising model systems. In this section we extend the Markov-property, Monte-Carlo investigation to two dimensions. To remind the reader briefly, the Markov prop-

Springer Proceedings in Physics, Vol. 78
Computer Simulation Studies in Condensed-Matter Physics VII
Editors: D.P. Landau · K.K. Mon · H.-B. Schüttler

erty means that for an Ising spin system, if we fix all the spins on the boundary of a finite region, then the expectation value of any function supported entirely in that region is completely independent of the spins outside that region. Baker [2] proposed the idea of dividing the plane-square lattice into diamond shaped regions, tabulating the sums of the appropriate quantities over all the interior spins of the diamonds as functions of the boundary-spin values, and then doing a Monte Carlo evaluation over the remaining boundary spins. Very significant speedups were obtained in one dimension [1].

In this study, we have considered four cases. The first case is meant to serve as a baseline and is the standard red-black decomposition of the lattice together with the standard Metropolis algorithm [3]. Since the purpose of this study is to consider the possible improvements due to block-internal spin summation, we use the straightforward Metropolis algorithm rather than a more sophisticated algorithm such as the Swendsen-Wang algorithm [4]. We used the Connection Machines routine "fast_rng" for the generation of the pseudo-random numbers. It is a lagged Fibonacci algorithm and we always used a 4 digit prime number as the initial seed. In this paper we use the term "auto-correlation time" to mean the number of "lattice updates" required so that the magnetization auto-correlation is equal to e^{-1} where e is the base of the natural logarithms. We denote this auto-correlation time by τ. By the term "lattice update" we mean that we apply the Metropolis algorithm once to each lattice spin to see if it should be flipped. In order to compute the auto-correlation time, we proceed as follows. Starting with a spin configuration, we perform a sufficient number, m, of lattice updates so that we estimate that the correlation between the magnetization at the original configuration and that in the new configuration to be about one-half. Typically our studies of the auto-correlation time were done with one sample taken every m lattice updates. Then 32 such sampled configurations were used in each of 40 coarse grain samples. These calculations were done on the CM-200. Some additional computations with 512 such sampled configurations for each of the 40 coarse grained samples were made on the CM-5. If the initial estimate and the observed auto-correlation times were too different, new estimates were used and the process continued until the estimate and the observations were fairly close. We have then computed the auto-correlation time by fitting a single exponential to the observed auto-correlation at a point where this auto-correlation is about one-half. This procedure yields estimates of the auto-correlation time which have statistical errors of about 2 - 3 percent which seems to be comparable to the accuracy claimed [5] for the same amount of computation for the sum of the observed auto-correlations over a window of time whose width is proportional to the auto-correlation time.

The second case studied was for the division of the spins of the plane-square lattice into diamonds with 2 spins on each edge. They have 4 boundary spins and one interior spin. As explained above, this process is equivalent to picking a Monte Carlo configuration of the boundary spins and then summing the internal spins in each block to obtain the statistical weight of the boundary spin configuration. A comparison of the flipped and non-flipped configuration weights by means of the standard, Metropolis, asymmetric algorithm allows us to decide whether to accept or reject the spin-flipped configuration.

The third case divides up the plane into diamonds with 3 spins on each edge so that there are 8 boundary spins and 5 interior spins. The fourth case divides the plane up into diamonds with 5 spins on each edge so that there are 16 boundary spins and 25 internal spins in each diamond. The case of 4 spins on the edge was skipped because this division of the plane has a repeat distance of 6 which is not a power of two.

We have performed these calculations for a grid of about 8 inverse temperatures ranging over $0.1K_c \leq K \leq 0.975K_c$, where for the plane-square Ising model $K_c = 0.440686794\ldots$. We have then fitted the auto-correlation times to a power law, with the most weight being given to the points closest to the critical temperature. We have obtained these results,

$$\begin{aligned}\tau_1 &\asymp 0.46(1 - K/K_c)^{-2.17},\\ \tau_2 &\asymp 0.22(1 - K/K_c)^{-2.05},\\ \tau_3 &\asymp 0.19(1 - K/K_c)^{-1.97},\\ \tau_5 &\asymp 0.137(1 - K/K_c)^{-1.90},\end{aligned} \tag{2.1}$$

where the subscripts 1, 2, 3, 5 refer to the four cases described above. The dynamical scaling exponent, z is defined by $\tau = \xi^z$, where ξ is the spin-spin, correlation length. Since in the two-dimensional Ising model the correlation length exponent ν is exactly unity, as $\xi \propto (1 - K/K_c)^{-\nu}$ the exponents in (2.1) are direct estimates for the dynamical critical exponents in each of the above methods. The exponent z describes the "critical slowing down" in the sense that the length of the necessary Monte Carlo simulations for a given accuracy is proportional to the auto-correlation time which in turn is governed by z. The simulation time increases dramatically as the critical point is approached when z is large. By way of comparison, Stauffer [6] quotes the Monte Carlo result $z_1 = 2.17 \pm 0.01$, and Dammann and Reger [7] give the series result $z_1 = 2.183 \pm 0.005$.

In order to estimate the effectiveness of these procedures, we have run sample problems on 32 nodes of the CM-5. Since each node has four vector units attached to it, this configuration is effectively equivalent to 128 central processing units (cpus). This is to be contrasted with the 512 cpus used in our computations on the CM-200. We have used a trial problem where the inverse temperature is $0.9K_c$ on a 64×64 lattice for the basic problem. The number of lattice updates between the sampled configurations is as given by (2.1), and we have used 512 sampled configurations for each of the 40 coarse grained samples, as explained above. For this problem we have considered a mode where we spread the problem across the machine and using all the nodes to process the problem in parallel, and the "idiot parallelization" mode, where 128 independent Monte Carlo problems (reducing the number of sample configurations per coarse grained step correspondingly) were run with each one running in a single cpu. For these parameters in cases 2 - 4, for the very same code, the idiot method was an order of magnitude faster on the CM-5 than the fully parallelized method, while in case 1 there was little time difference. For the trial problem we have found for case 1, 2381 seconds, for case 2, 1727, for case 3, 852, and for case 4, 267 seconds. Thus, the block-internal spin-summation method results in an order of magnitude improvement in the speed for the trial problem. The programs were all written in (Connection Machine) Fortran 90 in the same general style and no particular effort has been devoted to streamlining them as our current interest is in their relative speed.

A further set of parameters has been run for case 4 (the five-diamonds). A 256×256 lattice which has 16 times the number of spins in the trial problem was spread across the machine, but was run for only 32 repetitions for each of the 40 coarse grained samples. Since it was run for $K = 0.975K_c$, the number of lattice updates per sample is 151 instead of the 11 for the trial case. It required 7148 seconds which is roughly twice as long as would be expected in the idiot mode. Presumably for larger problems yet, the gap between the different modes will close further.

The computational times quoted in this section have been computed for comparability from actual cases which did not have exactly the proper number of skips between the sampled configurations.

3. Finite Size Behavior in Three Dimensions.

In a previous paper Baker [2] reported on the finite-size effects on the renormalized coupling-constant in two dimensions. The relevant thermodynamic function is,

$$g_r(K) = -\frac{v}{a^d}\frac{\dfrac{\partial^2\chi(K)}{\partial H^2}}{\chi^2(K)\xi^d(K)}, \tag{3.1}$$

$$g^* = \lim_{K\to K_c^-} g_r(K), \tag{3.2}$$

where g^* is the renormalized coupling constant for the model. The quantity v is the volume of a unit cell on the lattice, a is the lattice spacing, χ is the susceptibility, H is the magnetic field, and d is the spatial dimension. The series results imply [8] that $g^* = 14.66 \pm 0.06$. In that paper [2] it was found that in two dimensions the value of $\lim_{L\to\infty} g_r(K_c; L)$ was less than about 3.1 while the values of $g_r(K; L)$ obtained for $K < K_c$ are very consistent with the series results for an infinite system, provided $\xi_L/L \leq 1/(7 \pm 1)$, where L is the system width. This difference was explained by analysis of finite-size effects. The correlation length was determined by fitting as parameters ξ_L^{-2} and A in,

$$4\sum_{\omega=1}^{d}\sin^2(\frac{1}{2}\vec{q}\cdot\vec{e}_\omega)\left(1-\frac{\chi(L^d,\vec{q},K)}{\chi(L^d,\vec{0},K)}\right)^{-1} = \xi_L^{-2} + A\sum_{\omega=1}^{d}\sin^2(\frac{1}{2}\vec{q}\cdot\vec{e}_\omega), \tag{3.3}$$

to the five smallest, non-zero values of $\vec{q}$ allowed by periodic boundary conditions. Eq. (3.3) is exact with $A = 4$ for the one-dimensional Ising model.

It is of considerable interest to explore the finite-size behavior of $g_r(K; L)$ in three dimensions. To this end we have run a standard Monte Carlo code, (similar to that of Freedman and Baker [9]) on the CDC7600 for a very substantial period of time. Some of our results are displayed in Fig. 1, which shows

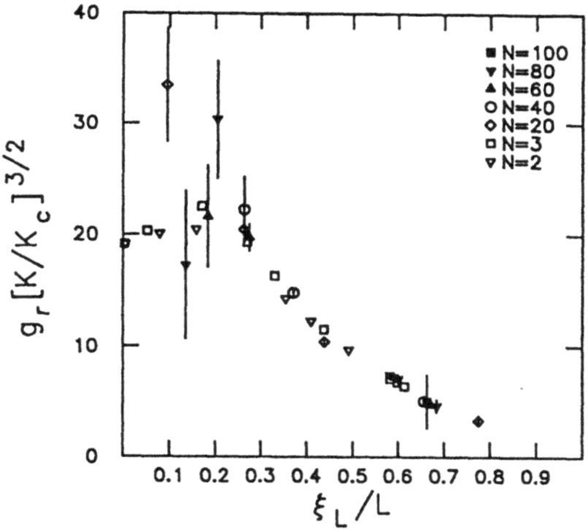

Fig. 1 Plot of $g_r(K;L)(K/K_c)^{3/2}$ for the three dimensional Ising model for simple cubic lattice systems of size $L \times L \times L$ spins versus $\xi_L(K)/L$. The point for $\xi_L = 0$ is common for all values of L.

the behavior of $g_r(K;L)$ when plotted against ξ_L/L. Results are given for $L = 2, 3, 20, 40, 60, 80, 100$. The results for $L = 2, 3$ are from exact summation and not Monte Carlo. The attention of the reader is drawn to the region of the plot where $\xi_L/L > \frac{1}{4}$. In this region we see very substantial data collapse, and it is in this region that we see that the finite-size effect is dominant. In the case of quantities which diverge, this finite-size effect is called finite-size rounding. In the simulation studies of the the three dimensional Ising model, it is also of interest to know how large a value of ξ_L/L can be used and still keep $|g_r(K;L)/g_r(K;\infty) - 1| < 0.01$, for example. {When K is not too close to K_c, $g_r(K;\infty)$ can be evaluated to adequate accuracy by series methods [8].} For L=2, the answer is $\xi_2/2 \approx 0.08$ and for L=3, the answer is $\xi_3/3 \approx 0.11$. For the case $L = 20$, $K = 0.212$, we compute that $\xi_{20} = 5.26 \pm 0.03$ and for the case $L = 40$, $K = 0.215$ we compute that $\xi_{40} = 4.46 \pm 0.09$. Since the correlation length in this region of K is monotonic in K, it must be that for $L = 20$ and $K = 0.212$ that the finite-size, correlation-length value is well over a percent off the infinite-system value. Note is made that series analysis gives $\xi(0.212) \approx 3.6$. Thus $\xi_L/L = 0.26$ is too large for accurate work. This result represents a criticism of the work of Freedman and Baker [9] who used $\xi_L/L \approx 0.275$, and also of the work of a number of other workers. Series analysis is consistent with the above quoted results for $K = 0.215$, $L = 40$ within the errors of both analysis. Since the results for $L = 2, 3$ suggest that the allowed value of ξ_L/L increases with L, we think that the allowed value of ξ_L/L for large systems is somewhere between 0.11 and 0.26. The values $\xi_L/L \leq 0.10$ may very likely be safe for one-percent-level work.

The authors wish to thank the Los Alamos Advanced Computing Laboratory for making time available on the CM-200 and the CM-5 for this project.

References

1. G. A. BAKER, JR., *A Markov-Property, Monte-Carlo Method: One Dimensional Ising Model*, J. Stat. Phys. **72** (1993), 621.
2. G. A. BAKER, JR., *The Markov-Property Method Applied to Ising Model Calculations*, Los Alamos preprint, submitted for publication.
3. N. METROPOLIS, A. W. ROSENBLUTH, M. N. ROSENBLUTH, A. H. TELLER, and E. TELLER, *Equation of State Calculations by Fast Computing Machines*, J. Chem. Phys. **21** (1953), 1087.
4. R. H. SWENDSEN and J.-S. WANG, *Nonuniversal Critical Dynamics in Monte Carlo Simulations*, Phys. Rev. Letts. **58** (1987), 86.
5. N. MADRAS and A. D. SOKAL, *The Pivot Algorithm: A Highly Efficient Monte Carlo Method for the Self-Avoiding Walk*, J. Stat. Phys. **50** (1988), 109.
6. D. STAUFFER, *Ising-Model Relaxation as Testing Ground for Modern Computers*, Computer Simulation Studies in Condensed-Matter Physics VI, D. P. Landau, K. K. Mon, and H.-B. Schüttler, eds., Springer-Verlag, 1993, Berlin.
7. B. DAMMANN and J. D. REGER, *Dynamical Critical Exponent of the Two Dimensional Ising Model*, Johannes Gutenberg-Universität Mainz preprint, (1992).
8. G. A. BAKER, JR., *Analysis of hyperscaling in the Ising Model by the high-temperature series method*, Phys. Rev. B **15** (1977), 1552-1559.
9. B. A. FREEDMAN and G. A. BAKER, JR., *A test for hyperscaling violation in the three-dimensional Ising model*, J. Phys. A **15** (1982), L715.

Decoupled-Cell Quantum Monte Carlo: How Should We Calculate the Energy?

C.J. Sisson

Department of Physics and Astronomy, Appalachian State University, Boone, NC 28608, USA

Abstract. An alternative and more general method for calculating the energy in a Decoupled Cell Monte Carlo simulation is presented. Examining this method on the one-dimensional XY model, the energy calculated from the cell energies is found to be well behaved in the low temperature limit, unlike the traditional energy calculation based on spin-spin correlations. In addition, the cell energies scale with the cell size, and thus can be extrapolated to infinite cell size to further remove errors arising from the finite size of the cell in the Monte Carlo.

1. Introduction

The Decoupled Cell Method (DCM) introduced by Homma et al. [1] is a relatively new approach to quantum Monte Carlo. DCM does not use the Trotter decomposition and therefore does not require simulation in an imaginary time direction. This has the advantage of requiring only a d-dimensional lattice to simulate a d-dimensional system, and makes possible DCM simulations of quantum systems that are much larger than when the Suzuki-Trotter approach is used. DCM has been applied to quantum spin systems ranging from the XY and XXZ models to the Heisenberg model [1 – 6].

The traditional method of calculating the energy in a DCM simulation makes use of a spin-spin correlation function. While this method performs very well in the medium and high temperature regions, it breaks down in the low temperature limit. In addition, this method of calculating the energy is not easy to generalize to models that have contributions from an on-site potential energy or kinetic energy.

In this paper an alternative method for calculating the energy within a DCM simulation is presented.

2. The Decoupled Cell Method

Unlike the time-like decomposition of a quantum Hamiltonian used in the Suzuki-Trotter method for quantum Monte Carlo, the Decoupled Cell Method uses a space-like decomposition. The approximation used in a DCM simulation is that a quantum Hamiltonian for the entire system H can be decomposed into a series of smaller cell Hamiltonians h_i: $H = h_1 + h_2 + \dots$. This allows one to approximate the Boltzmann factor as

$$e^{-\beta H} \approx \prod_i e^{-\beta h_i} \quad ,$$

where β is the inverse temperature $\beta = 1/k_B T$ and the cell Hamiltonians are as large as possible.

Springer Proceedings in Physics, Vol. 78
Computer Simulation Studies in Condensed-Matter Physics VII
Editors: D.P. Landau · K.K. Mon · H.-B. Schüttler

Under this assumption, the probability of flipping a single spin s_i is given by

$$P(\psi(s_i) \rightarrow \psi(-s_i)) \approx \frac{\langle\psi(-s_i)|\prod_j e^{-\beta h_j}|\psi(-s_i)\rangle}{\langle\psi(s_i)|\prod_j e^{-\beta h_j}|\psi(s_i)\rangle}$$
$$\approx \frac{\langle\psi(-s_i)|e^{-\beta h_i}|\psi(-s_i)\rangle}{\langle\psi(s_i)|e^{-\beta h_i}|\psi(s_i)\rangle} \quad ,$$

as all the terms not containing the site i drop out. Once the eigenvalues (ϵ_n) and eigenvectors ($|n\rangle$) for the cell Hamiltonians are known, the (unnormalized) probability for finding the system in a state ψ_α is given by

$$P(\psi_\alpha) = \langle\psi_\alpha|e^{-\beta h_i}|\psi_\alpha\rangle = \sum_n |\langle\psi_\alpha|n\rangle|^2 e^{-\beta\epsilon_n} \quad .$$

For a more complete description of the method see any of the papers by Homma et al. [1—4, and references therein], or by Creswick and Sisson [5,6].

DCM has been used primarily on interacting spin systems such as the XY, XXZ and Heisenberg models. For these models, the energy can be calculated from a spin correlation function. In the XY model, for example, the Hamiltonian is given by

$$H = -2J\sum_{\langle ij\rangle}(S_i^x S_j^x + S_i^y S_j^y) \quad ,$$

and the energy can be calculated as

$$E = \langle H\rangle = -2J\sum_{\langle ij\rangle}\langle S_i^x S_j^x\rangle + \langle S_i^y S_j^y\rangle = -4J\sum_{\langle ij\rangle}\langle S_i^x S_j^x\rangle \ ,$$

where the final step assumes that there is no spontaneous symmetry breaking ($\langle S_i^x S_j^x\rangle = \langle S_i^y S_j^y\rangle$).

However, it is not clear how one can generalize this to models with energy contributions from electron motion and on-site potentials. This excludes several important models in condensed matter physics, including the t-J and Hubbard models.

As an alternative to the correlation function approach to the energy, one can calculate "cell energies" in a particular state in analogy with the cell probability for a particular state. This uses a statistical mechanics approach by finding the energy from the logarithmic derivative of the decoupled cell partition function:

$$\langle E\rangle = -\frac{\partial \ln Z}{\partial\beta}$$
$$= -\frac{\partial}{\partial\beta}\ln\{\sum_\alpha\langle\psi_\alpha|\prod_i e^{-\beta h_i}|\psi_\alpha\rangle\} \quad .$$

Working through the derivative, one can simplify the final expression for the average energy by defining the cell energy $e(\psi_{\alpha(i)})$ to be

$$e_i(\psi_{\alpha(i)}) \equiv \frac{\langle\psi_{\alpha(i)}|h_i e^{-\beta h_i}|\psi_{\alpha(i)}\rangle}{\langle\psi_{\alpha(i)}|e^{-\beta h_i}|\psi_{\alpha(i)}\rangle}$$

and the probability for finding the lattice in a state $|\psi_\alpha\rangle$ to be

$$P(\psi_\alpha) \equiv \prod_i\langle\psi_{\alpha(i)}|e^{-\beta h_i}|\psi_{\alpha(i)}\rangle \quad .$$

This gives an expression for the average energy of

$$\langle E\rangle = \frac{1}{Z}\sum_{s} P(\psi_\alpha)\sum_{i} e_i(\psi_{\alpha(i)}) \quad .$$

In practice, the cell energies for each cell state $e(\psi_{\alpha(i)})$ can be calculated before the Monte Carlo simulation.

3. Results

This method was applied to the 1 dimensional XY model on a 150 site lattice, for cells ranging in size from 3 sites to 11 sites. All systems were equilibrated for 50,000 Monte Carlo steps per site and then averages were calculated over an additional 25,000 MCS/site. Temperatures ranged from $1.5J$ to $0.1J$.

Figure 1 compares the energy calculated in these simulations with the exact (infinite lattice) solution to the XY model given by Katsura [7]. The energies are well behaved in the low temperature limit for all cell sizes and follow the general shape of the exact solution. Additionally, there is scaling with the size of the cell used — all simulations took place on 150 site lattices, so there is no lattice scaling here. In fact, as discussed in references [8] and [9], the energies scale with the number of bonds in the cell for this model.

Extrapolating these energies against the inverse number of bonds on to the infinite cell limit yields good agreement with Katsura's exact solution. However, the largest discrepancies from the scaling occur with the smallest cells. Best results can be obtained by using only the two largest cells in the extrapolation. Figure 2 shows the results when the 9-site and 11-site cells are used: the energies are within 1.0% of the exact solution for temperatures down to $0.2J$. Using the 7-site and 9-site cells gives only slightly worse results, with agreement within 1.5% of the exact result for the same range of temperatures. At the very lowest temperatures the 9-site and 11-site energies cross, so that although the energies for the individual cells are well behaved, the extrapolated

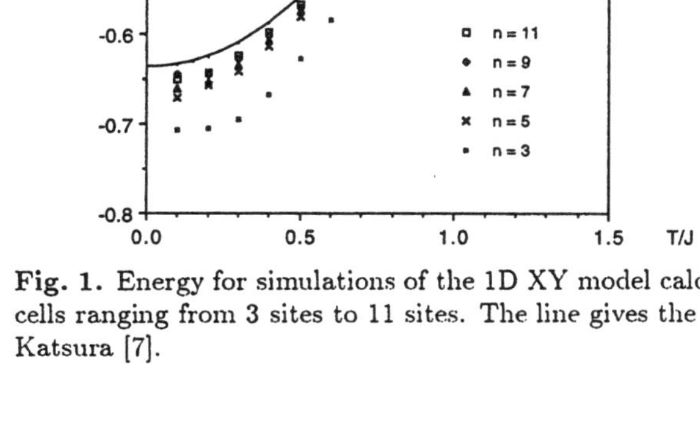

Fig. 1. Energy for simulations of the 1D XY model calculated using the cell energy for cells ranging from 3 sites to 11 sites. The line gives the exact (infinite lattice) result of Katsura [7].

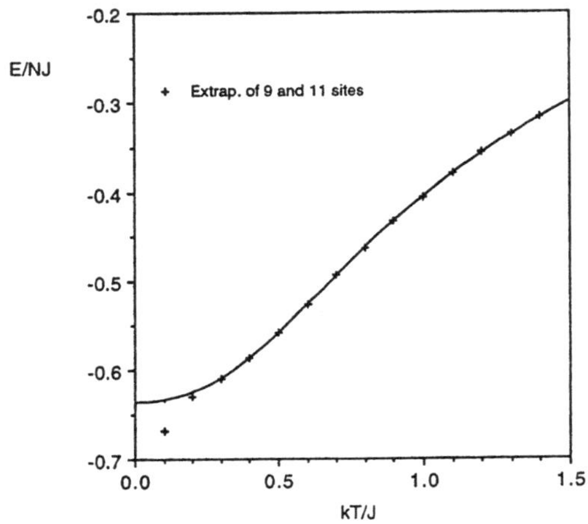

Fig. 2. Energy extrapolated from the 9 and 11 site cells. Values are within 1.5% of the exact value for temperatures down to $1.5J$.

energy diverges from the exact result. DCM is not an appropriate choice for very low temperature simulations regardless of what method is used for calculating the energy. By using the cell energies, however, one can examine lower temperatures than otherwise possible.

References

1. S. Homma, H. Matsuda, and N. Ogita, Prog. Theor. Phys. **72** (1984), 1245.
2. S. Homma, H. Matsuda, and N. Ogita, Prog. Theor. Phys. **75** (1986), 1058.
3. S. Homma, K. Sano, H. Matsuda, and N. Ogita, Prog. Theor. Phys. Supp. **87** (1986), 127.
4. S. Homma, H. Matsuda, N. Ogita, and K. Sano, J. Phys. Soc. Jpn. **62** (1993), 880.
5. R. Creswick and C. Sisson, Mod. Phys. Lett. B **5** (1991), 907.
6. R. Creswick and C. Sisson, In *Quantum Monte Carlo Methods in Condensed Matter Physics* ed. M. Suzuki, (World Scientific, 1993).
7. S. Katsura, Phys. Rev. **127** (1962), 1508.
8. C. Sisson, Ph.D. dissertation (1993).
9. C. Sisson and R. Creswick, preprint.

Computer Study of a $d = 2$ Mixed Ising Ferrimagnet

G.M. Buendıa, M.A. Novotny, and J. Zhang*

Supercomputer Computations Research Institute,
Florida State University, Tallahassee, FL 32306-4052, USA
*Permanent address: Department of Physics,
Universidad Simón Bolivar, Caracas 1080, Venezuela

Abstract. A model for a square-lattice ferrimagnet is investigated using ground-state analysis, Monte Carlo simulations, and numerical transfer-matrix studies. The model has spin $\frac{1}{2}$ on one sublattice and spin 1 on the other sublattice, and all possible nearest-neighbor and field interactions. Our preliminary results indicate an Ising-like transition with no compensation point and no tricritical point, in disagreement with a mean-field analysis.

1. Introduction

Although the theoretical studies of ferrimagnetic systems have been performed since the time of Néel [1,2], there is recent experimental motivation for reopening this area of research. The experimental motivation is the strong and continuing effort to obtain low-density, transparent magnets with spontaneous moments at room temperature [3]. Synthesis of single-chain and double-chain ferrimagnets are now becoming standard [4]. However, attempts to synthesize higher dimensional low-density ferrimagnets are only now coming to fruition. Example materials are the recently developed amorphous $\mathrm{V(TCNE)}_x \cdot y(\mathrm{solvent})$ materials which show ordering temperatures as high as 400 K [5], 3d ferrimagnets with T_c up to 240 K [6], a material which has not yet been crystalized and which may be a d=2 ferrimagnet [7], as well as other d=2 organometalic ferrimagnets [8].

We will study a square-lattice mixed Ising model that may be applicable to d=2 ferrimagnets. One question of interest will be the phase diagram. Another will be the possible existence of a compensation point (CP). A CP occurs in ferrimagnets of type N as classified in Ref. [1]. The CP is the one temperature where the resultant magnetization vanishes below the critical point. The behavior near the compensation point may be of technological significance since at this point only a small driving field is required to change the resultant magnetization of the material. It has been shown [9] via exact mappings that the square-lattice spin $\frac{1}{2}$ antiferromagnet with different fields applied to the two sublattices has no CP, while a decorated square lattice does have a CP.

We are interested in mixed Ising systems, since in these materials behavior of type N may occur. The Hamiltonian we study includes all possible nearest-neighbor (nn) interactions and fields. It is given by

$$\mathcal{H} = -J_1 \sum_{\langle i,j \rangle} \sigma_j S_i - J_2 \sum_{\langle i,j \rangle} \sigma_j S_i^2 - H_{\frac{1}{2}} \sum_j \sigma_j - H_1 \sum_i S_i + D \sum_i S_i^2, \quad (1)$$

Springer Proceedings in Physics, Vol. 78
Computer Simulation Studies in Condensed-Matter Physics VII
Editors: D.P. Landau · K.K. Mon · H.-B. Schüttler

where the first two sums are over all nn spin pairs, and the spin variables on the two sublattices are chosen to be $\sigma_j \in [-1,+1]$ and $S_i \in [-1,0,+1]$, respectively. The model with $J_2=H_1=H_{1/2}=0$ has been studied by a mean-field analysis [10], and a tricritical point and CP were found.

2. Ground-State Diagram

We use a 2×2 cell, which has $2^2 3^2/2$ configurations, with the division by 2 due to rotational symmetry. These 18 configurations are shown in Fig. 1. According to the Hamiltonian, Eq. (1), the energies of the 18 ground-state configurations are the following,

1. $-H_{1/2}/2$	10. $-H_{1/2}/2+D/2-2J_2$
2. 0	11. $D/2$
3. $H_{1/2}/2$	12. $H_{1/2}/2+D/2+2J_2$
4. $-H_{1/2}/2-H_1/4+D/4-J_1-J_2$	13. $-H_{1/2}/2-H_1/2+D/2-2J_1-2J_2$
5. $-H_1/4+D/4$	14. $-H_1/2+D/2$
6. $H_{1/2}/2-H_1/4+D/4+J_1+J_2$	15. $H_{1/2}/2-H_1/2+D/2+2J_1+2J_2$
7. $-H_{1/2}/2-H_1/4+D/4+J_1-J_2$	16. $-H_{1/2}/2+H_1/2+D/2+2J_1-2J_2$
8. $H_1/4+D/4$	17. $H_1/2+D/2$
9. $H_{1/2}/2+H_1/4+D/4-J_1+J_2$	18. $H_{1/2}/2+H_1/2+D/2-2J_1+2J_2$

Restricting ourselves to $H_{1/2}=H_1=0$ gives a ground-state diagram in which the favorable configurations are the disordered configurations (1st, 2nd, and 3rd) and the 15th and 16th configurations, as labelled in Fig. 1 . The ground-state

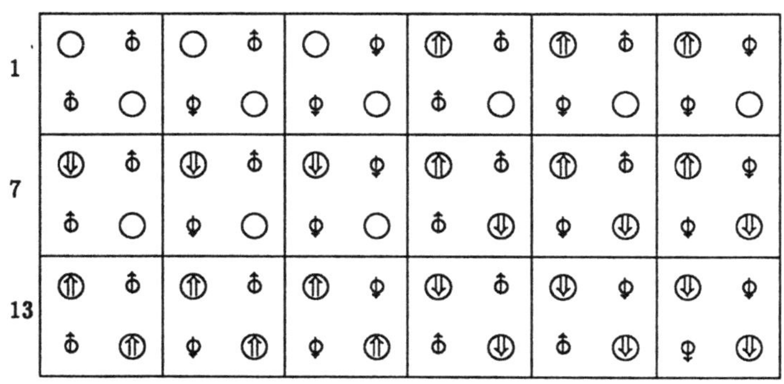

Fig. 1 The 18 possible spin configurations in the unit cell are shown. The large circles are the sites with spin 1, and the smaller circles are the sites with spin $\frac{1}{2}$. The arrows correspond to spins +1 (up arrow), −1 (down arrow), and 0 (empty large circle). The unit cells are numbered from left to right, with the numbering of the left-most cells shown at left.

plane, ($x{=}D/|J_1|$,$y{=}J_2/|J_1|$), is separated into 3 regions. A line along the x axis from $-\infty$ to 4 separates the regions with the ground-state configurations labeled 15 and 16. Two lines given by $y{=}\pm(x/4)\mp1$ for $x{\geq}4$ separate the 16th and 15th configuration regions from the region of the disordered configurations.

3. Monte Carlo Calculations

We applied standard importance sampling methods to perform Monte Carlo calculations with the Hamiltonian of Eq. (1). Periodic boundary conditions on $L{\times}L$ lattices were used. Configurations were generated by sequentially traversing the lattice and making single spin flip attempts. The flips are accepted or rejected with standard heat-bath dynamics. Data were generated with 10^4 to 10^5 Monte Carlo steps per site, and the error bars were taken from the standard deviation of blocks of 10^2 measurements each. Our program calculates the internal energy, the specific heat, the total and sublattice magnetizations, two order parameters, and the associated susceptibilities. We defined one order

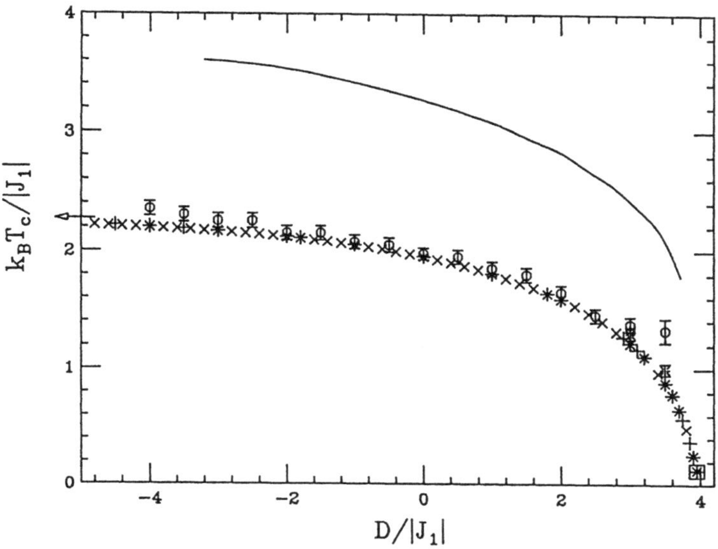

Fig. 2 The phase diagram for $H_1{=}0$, $H_{\frac{1}{2}}{=}0$, and $J_2{=}0$ is shown as function of $D/|J_1|$. The solid line corresponds to the mean-field calculation [10]. The circles (stars) with error estimates are Monte Carlo results for the location of the specific heat maximum on a lattice with $L{=}10$ ($L{=}16$). Numerical transfer-matrix calculations are also shown for system sizes $N{\times}\infty$ and $(N{+}1){\times}\infty$ with $N{=}4$ ($\times$) and $N{=}6$ ($+$) and $N{=}8$ (square). In the limit $D{\rightarrow}{-}\infty$ the model reduces to a nn spin 1/2 model, and the arrow marks the exact value of T_c for this limit. Excellent agreement is seen between the Monte Carlo and numerical transfer-matrix results.

parameter as

$$O_1 = \frac{1}{L^2}\left\langle\left|\sum_{i,j} S_i + \frac{1}{2}\sigma_j\right|\right\rangle, \tag{2}$$

where the sums are over all sites and the averages are over all configurations. Any CP should be seen by O_1. We checked that our results are in agreement with exact-enumeration studies for $L=2$, and that the ground-state diagrams are reproduced for the different combinations of parameters in the Hamiltonian.

In Fig. 2 we show the dependence of the critical temperature on D with the parameter $J_1=-1$. The values were obtained by locating the maximum value of the specific heat. Also, we show the mean-field prediction for this model [10]. We found reasonable agreement between the general shape of both curves. However, the mean-field values are substantially higher. We observed large finite-size effects near the special point $(x,y)=(4,0)$ in the ground-state diagram.

So far we have been unable to find any evidence for the existence of a compensation point in the region of the J_1–D plane where mean-field theory [10] predicts it to be.

4. Transfer-Matrix Calculations

Traditional numerical transfer-matrix calculations were performed [11]. The transfer direction on the square lattice was chosen to be in the (11) direction. For a strip with N spins the symmetric transfer matrix is given by $\mathbf{T}=\mathbf{D}_{1/2}^{1/2}\mathbf{A}^{\mathrm{T}}\mathbf{D}_1\ \mathbf{A}\mathbf{D}_{1/2}^{1/2}$ where $\mathbf{A}$ is a $3^N\times 2^N$ matrix that includes the nn interactions J_1 and J_2 between layers. The diagonal matrix $\mathbf{D}_{1/2}$ ($\mathbf{D}_1$) is of size $2^N\times 2^N$ ($3^N\times 3^N$) and contains interactions within the layers given by $H_{1/2}$ (D and H_1). The numerical diagonalization was performed using the NAG routine F02FJE to find the two largest eigenvalues λ_i.

The ratio of the largest and next-largest eigenvalues of $\mathbf{T}$ gives the correlation length $\xi_N^{-1} = \ln|\lambda_1/\lambda_2|$. The scaling form for ξ is $\xi_N=N\mathcal{F}(tN^{y_T})$ where $t = |(T-T_c)/T_c|$ with T_c the critical temperature [11]. An estimate for T_c is given by the phenomenological scaling relation $\xi_{N_1}/N_1=\xi_{N_2}/N_2$ [11]. Differentiating the scaling relation at this estimate for T_c gives the finite-strip-width estimate of the critical exponent $y_T=1/\nu$ [11].

Fig. 2 shows results using $N\times\infty$ and $(N+1)\times\infty$ lattices with $N=4$, 6, and 8. The results are in excellent agreement with our Monte Carlo calculations. The finite-strip width estimates for y_T were consistent with the Ising value $y_T=1$ (to at least within 3%) for all D values shown in Fig. 2. Indeed, as N increases the values for y_T approach the Ising value for all $x<4$. This indicates that there is no tricritical point at finite temperatures. The numerical transfer-matrix results only indicated a finite-temperature phase transition for $x<4$ and $y=0$. Thus these numerical transfer-matrix results indicate that there is a multicritical point only at $T=0$, located at $(x,y)=(4,0)$.

5. Discussion and Conclusions

We have introduced a square-lattice mixed Ising model for d=2 ferrimagnets. We have performed a ground-state analysis, Monte Carlo simulations, and numerical transfer-matrix calculations on this model. A mean-field analysis with only J_1 and D not equal to zero found a tricritical point and a range of D values with a compensation point [10]. Our preliminary results do not find any indication of a tricritical point at finite temperatures, and no indication of a compensation point for any D and J_2. Consequently, if a compensation point is found experimentally in d=2, a more complicated Hamiltonian would be required to model the material. This could be accomplished, for example, by including next-nearest-neighbor interactions.

Acknowledgements

Supported in part by the Florida State University Supercomputer Computations Research Institute, which is partially funded through Contract #DE-FC05-85ER25000 by the U.S. Department of Energy, and by Florida State University through time granted on its CRAY Y-MP supercomputer. G. M. B. also acknowledges financial support by the Venezuelan Science Counsel, CONICIT. Useful discussions with C. P. Landee, P.-A. Lindgård and P. A. Rikvold are acknowledged.

References

[1] L. Néel, Ann. de Phys. **3**, 137 (1948).
[2] B. Szpunar and P.-A. Lindgård, Phys. Stat. Sol. B **82**, 449 (1977).
[3] *NATO ASI Series*, edited by D. Gatteschi, O. Kahn, J. S. Miller, and F. Palacio (Kluwer Academic, Dordrecht, 1991).
[4] For a theoretical investigation see: J. Curély, R. Georges, M. Drillon, M. Belaiche, and K. Benkhouja, Phys. Rev. B **46**, 3527 (1992).
[5] G. Du, J. Joo, and A. J. Epstein, J. Appl. Phys. **73**, 6566 (1993).
[6] T. Mallah, S. Thiébaut, M. Verdaguer, and P. Veillet, Science **262**, 1554 (1994).
[7] H. Tamaki, Z. J. Zhong, N. Matsumoto, S. Kida, M. Koikawa, N. Achiwa, Y. Hashimoto, and H. Okawa, J. Am. Chem. Soc. **114**, 6974 (1992).
[8] C. P. Landee and M. M. Turnbull, private communication.
[9] I. Syozi and H. Nakano, Prog. Theor. Phys. **13**, 69 (1955).
[10] T. Kaneyoshi and J. C. Chen, J. Magn. Mag. Mat. **98**, 201 (1991).
[11] M. P. Nightingale, in *Finite Size Scaling and Numerical Simulation of Statistical Systems*, ed. V. Privman (World Scientific, Singapore, 1990).

Computational Studies of Organolithium Compounds

L.M. Pratt and I.M. Khan
Department of Chemistry, Clark Atlanta University, Atlanta, GA, USA

Computational studies were performed on several representative types of organolithium compounds. Comparisons were made between the results using ab initio, density functional, and semiempirical methods. Both the ab initio and density functional methods generated similar bond lengths and dimerization energies for most compounds studied, while the MNDO calculations gave results which varied considerably with the type of compound studied.

1. Introduction

Organolithium compounds are among the most important reagents for carbon-carbon bond formation in organic synthesis. Sterically hindered lithium amide bases are used to selectively generate one of two possible enolate isomers in aldol type reactions. The selectivity is often dependent on the reaction conditions, and may be changed by the addition of inorganic lithium salts. [1] Butyllithium is commonly used as an initiator in anionic polymerizations, and the propagating chain end is also an alkyllithium species. Teyssie and coworkers have shown that the living character of some anionic polymerizations can be perturbed by the addition of lithium salts. [2] It is now known that most alkyllithiums, lithium amides, lithium enolates, and other lithium compounds exist as aggregates in most organic solvents.

Computational methods have become an important tool for the study of organolithium compounds. Ab initio calculations have been used in the study of several small organolithium molecules. We have recently reported the good performance of density functional theory (DFT) calculations on similar molecules. [3] Although these calculations slightly underestimate the lithium-carbon and lithium-nitrogen bond lengths, the error is smaller than that of semiempirical calculations for many types of compounds. The semiempirical MNDO [4] method has been shown to be in qualitative agreement with the experimental geometries and relative energies of solvated lithium amide aggregates. [5] Unfortunately, MNDO severely underestimates the repulsion between the lithium core electrons and the carbon lone pair electrons, which results in overestimation of the lithium-carbon bond strength and incorrect calculated geometries of many molecules which contain lithium-carbon bonds. [6] In this work we compare the three computational methods for several representative organolithium compounds.

Springer Proceedings in Physics, Vol. 78
Computer Simulation Studies in Condensed-Matter Physics VII
Editors: D.P. Landau · K.K. Mon · H.-B. Schüttler

2. Computational Methods

Ab initio calculations were performed using the Gaussian series of programs [7] on a Silicon Graphics Jurassic Classic workstation. Unless otherwise specified, the 6-311G* basis set was used for all calculations. Density functional calculations were performed with the DMol program and the Insight II graphical interface, both produced by Biosym, [8] on a Silicon Graphics Indigo 2 workstation. All DFT calculations were performed using a double numeric basis set with polarization functions (DNP), after preoptimization of the geometry with a minimal basis set. MNDO calculations were performed using the MOPAC [9] module of the Biosym Discover program.

3. Results and Discussion

Bond lengths and angles were calculated for monomeric (**1a**) and dimeric (**1b**) methyllithium by Gaussian, DMol, and MNDO. The ab initio and DFT calculations generated carbon-lithium bond lengths of 1.987 and 1.965 A, respectively, while the MNDO calculation resulted in a much shorter bond length of 1.821A. The results are summarized in Table 1.

Table 1. Calculated geometries of the methyllithium monomer

Method	C-Li	C-H	Li-C-H
MNDO	1.821	1.117	112.0
Ab Initio	1.965	1.111	113.3
DFT	1.987	1.094	113.2

The MNDO geometry of the methyllithium dimer was much closer to the ab initio and DFT geometries compared to the monomer. This appears to be due to the orientation of the lithium atoms relative to the carbon sp^3 orbital, which partially avoids the underestimation of the repulsion between the lithium core orbitals and the carbon lone pair as described above. The MNDO calculations greatly overestimated the dimerization energies of methyllithium. The results are summarized in Table 2.

Table 2. Calculated energies and geometries of the methyllithium dimer

Method	C-Li	C-Li-C	Li-C-Li	Dimerization Energy (kcal/mole)
MNDO	2.043	120.5	59.0	-72.2
Ab Initio	2.126	119.0	61.0	-43.0
DFT	2.089	116.8	63.1	-48.9

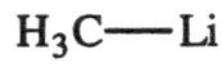 **1a** 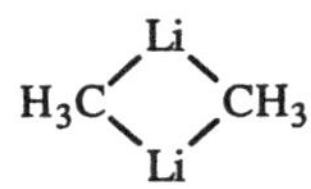 **1b**

 **2a** 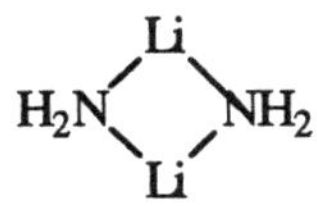 **2b** 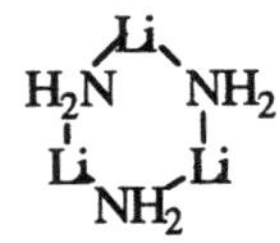 **2c**

The geometry of lithium amide was calculated by each of the three methods for the monomeric (2a), dimeric (2b), and trimeric (2c) forms. The calculated lithium-nitrogen bond distances were similar in the monomer, while MNDO overestimated this bond length in the dimer and underestimated it in the trimeric form. The results are summarized in Table 3.

Table 3. Lithium-nitrogen bond lengths in lithium amide aggregates

Method	Monomer	Dimer	Trimer
MNDO	1.753	2.034	1.541
Ab Initio	1.728	1.925	1.925
DFT	1.731	1.903	1.896

Several types of mixed aggregates have appeared in the literature. Lithium tetramethylpiperdide forms a 1:1 mixed dimer with ethyllithium, phenyllithium, and t-butyllithium, as well as a second aggregate with t-butyllithium of undetermined structure. [10] Mixed aggregates of unknown structure between lithium chloride and growing polymer chain ends has reportedly affected the living character of methacrylate polymerizations. [2] The mixed aggregates between lithium tetramethylpiperdide and lithium chloride has been studied by ^{6}Li and ^{15}N NMR spectroscopy. Calculations were performed on the 1:1 mixed dimer of lithium dimethylamide and lithium chloride (3) as a model system for this type of aggregate. The three methods yielded similar geometries, except that MNDO overestimated the Li-Cl bond length by about 0.3 A. The results are summarized in Table 4.

Table 4. Calculated geometry of the lithium dimethylamide-LiCl mixed dimer

Method	N-Li	Li-Cl	Li-N-Li	Li-Cl-Li
MNDO	2.036	2.504	80.1	63.1
Ab Initio	1.924	2.227	77.8	65.8
DFT	1.898	2.197	78.4	66.2

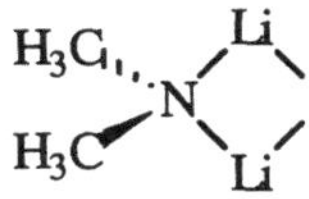

3

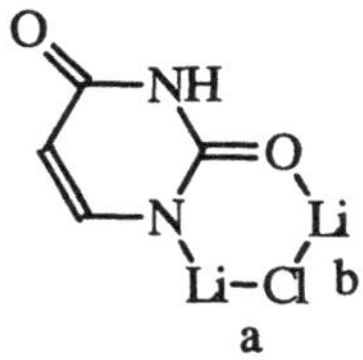

4

A similar mixed aggregate has been described between lithium chloride and lithiated uracil (4). [11] Due to the size of this system, ab initio calculations were performed using the 3-21G basis set, and DFT calculations with a minimal basis set. Both the ab initio and DFT calculations resulted in nearly equal Li-Cl bond lengths, while the MNDO calculation resulted in a bond length difference of 0.5 A. MNDO also greatly overestimated the lithium-oxygen bond distance. The results are summarized in Table 5.

Table 5. Bond distances in the lithiated uracil-LiCl mixed dimer

Method	N-Li	Li-Cl (a)	Li-Cl (b)	Li-O
MNDO	2.042	1.961	2.471	2.479
Ab Initio	1.906	2.293	2.284	1.704
DFT	1.823	2.182	2.200	1.751

We have reported the results of MNDO and DFT calculations on allyllithium and cyclopentadienyllithium. [3] In both cases the MNDO results were in fairly good agreement with the DFT results and the published ab initio data. [12-14] The relatively good MNDO geometries appear to be due to the π orbital system, which avoids overlap between the lithium atom and carbon lone electron pairs, as in the case of the methyllithium dimer described earlier.

4. Conclusions

We have examined three computational methods for the study of organolithium compounds. Ab initio and DFT calculations give consistent results for a wide variety of lithium compounds. The DFT calculations did underestimate the lithium-carbon and lithium-nitrogen bond lengths by about 2-5%, which is consistent with the known behavior of DFT in other small molecules. [15] Due to the lower computational requirements, DFT is able to handle somewhat larger systems than Hartree-Fock ab initio calculations. MNDO alone is capable of calculations of large solvated systems at this time. Due to the artifacts of the parameterization, MNDO gives incorrect geometries and energies for many compounds which contain carbon-lithium single bonds. MNDO does produce good qualitative geometries for compounds in which the overlap between lithium and carbon sp^3 orbitals is avoided. The results must be evaluated on a case by case basis, however, and should be compared to experimental and/or ab initio data whenever possible.

5. Acknowledgments

The authors gratefully acknowledge the support of this work by NSF Grant # RII-9005196 and NIH Grant # GM 08247-06. The use of the Gaussian program and workstation was provided by the Center for Theoretical Studies of Physical Systems (CTSPS), Clark Atlanta University. Purchase of the Silicon Graphics Indigo 2 workstation and the Biosym computational program by NASA Grant # NAGW-2939 is also acknowledged.

References

1. a. P. L. Hall, J. H. Gilchrist, D. B. Collum J. Am. Chem. Soc. 1991, 113, 9571-4

b. P. L. Hall, J. H. Gilchrist, A. T. Harrison, D. J. Fuller, D. B. Collum J. Am. Chem. Soc. 1991, 113, 9575-85

2. J. Wang, R. Jerome, R. Warin, P. Teyssie Macromolecules 1993, 26, 5984-90
3. L. M. Pratt and I. M. Khan, Submitted for publication
4. M. J. S. Dewar and W. J. Thiel J. Am. Chem. Soc. 1977, 99, 4899-4907
5. F. E. Romesberg and D. B. Collum J. Am. Chem. Soc. 1992, 114, 2112-21
6. R. Glaser and A. Streitwieser Theochem 1988, 163, 19-50
7. M. J. Frisch, G. W. Trucks, M. Head-Gordon, P. M. W. Gill, M. W. Wong, J. B. Foresman, B. G. Johnson, H. B. Schlegel, M. A. Robb, E. S. Replogle, R. Gomperts, J. L. Andres, K. Raghavachari, J. S. Binkley, C. Gonzalez, R. L. Martin, D. J. Fox, D. J. Defrees, J. Baker, J. J. P. Stewart, J. A. Pople GAUSSIAN 92, Gaussian, Inc. Pittsburgh, PA, 1992
8. Computational results obtained using software program from Biosym Technologies of San Diego, CA
9. J. J. P. Stewart QCPE 581
10. L. M. Pratt and D. B. Collum Unpublished results
11. L. M. Pratt and C. C. Chu A Computational Study of Structure and Bonding in Aggregates of Lithiated Uracil Theochem, In press
12. G. Decher and G. Boche J. Organomet. Chem. 1983, 259, 31-6
13. T. Clark, E. D. Jemmis, P. v. R. Schleyer, J. S. Brinkley, J. A. Pople J. Organomet. Chem. 1978, 150, 1-6
14. K. C. Waterman and A. Streitwieser J. Am. Chem. Soc. 1984, 106, 3138-40
15. J. Andzelm and E. Wimmer J. Chem. Phys. 1992, 96, 1280-1303

Index of Contributors

Amar, J.G. 57
Assaad, F.F. 113

Baker, G.A. Jr. 213
Bassler, K.E. 171
Berg, B.A. 18
Buendia, G.M. 223

Canright, G.S. 204
Choi, M.Y. 193

Dieterich, P. 113
Dünweg, B. 177

Erpenbeck, J.J. 213

Family, F. 57
Freericks, J.K. 99

Gerling, R.W. 145
Gorman, B.M. 199
Gunton, J.D. 7
Günther, C.C.A. 199

Hanke, W. 113
Hansmann, U.H.E. 183

Ihm, M.-O. 188
Ito, N. 44

Janke, W. 29
Jarrell, M. 99

Khan, I.M. 228
Kikuchi, M. 44

Lallemand, P. 209
Lam, P.M. 57
Landau, D.P. 1
Lee, J. 193
Linden, W. von der 113
Lindgård, P.A. 69

Mon, K.K. 1
Muramatsu, A. 113

Nielaba, P. 188
Noack, R.M. 85
Novotny, M.A. 161, 223

Okabe, Y. 44
Okamoto, Y. 183
Ortiz, G. 125

Pandey, R.B. 166
Pratt, L.M. 228
Preuss, R. 113

Scalapino, D.J. 85
Schneider, F. 188
Schüttler, H.-B. 1
Sisson, C.J. 219

Viñals, J. 7

Watson, G.I. 204
White, S.R. 85

Xi, H.W. 7

Zhang, J. 223

Springer Proceedings in Physics

Managing Editor: H. K. V. Lotsch

1 *Fluctuations and Sensitivity in Nonequilibrium Systems*
Editors: W. Horsthemke and D. K. Kondepudi

2 *EXAFS and Near Edge Structure III*
Editors: K. O. Hodgson, B. Hedman, and J. E. Penner-Hahn

3 *Nonlinear Phenomena in Physics*
Editor: F. Claro

4 *Time-Resolved Vibrational Spectroscopy*
Editors: A. Laubereau and M. Stockburger

5 *Physics of Finely Divided Matter*
Editors: N. Boccara and M. Daoud

6 *Aerogels* Editor: J. Fricke

7 *Nonlinear Optics: Materials and Devices*
Editors: C. Flytzanis and J. L. Oudar

8 *Optical Bistability III*
Editors: H. M. Gibbs, P. Mandel, N. Peyghambarian, and S. D. Smith

9 *Ion Formation from Organic Solids (IFOS III)*
Editor: A. Benninghoven

10 *Atomic Transport and Defects in Metals by Neutron Scattering*
Editors: C. Janot, W. Petry, D. Richter, and T. Springer

11 *Biophysical Effects of Steady Magnetic Fields*
Editors: G. Maret, I. Kiepenheuer, and N. Boccara

12 *Quantum Optics IV*
Editors: J. D. Harvey and D. F. Walls

13 *The Physics and Fabrication of Microstructures and Microdevices*
Editors: M. J. Kelly and C. Weisbuch

14 *Magnetic Properties of Low-Dimensional Systems*
Editors: L. M. Falicov and J. L. Morán-López

15 *Gas Flow and Chemical Lasers*
Editor: S. Rosenwaks

16 *Photons and Continuum States of Atoms and Molecules*
Editors: N. K. Rahman, C. Guidotti, and M. Allegrini

17 *Quantum Aspects of Molecular Motions in Solids*
Editors: A. Heidemann, A. Magerl, M. Prager, D. Richter, and T. Springer

18 *Electro-optic and Photorefractive Materials*
Editor: P. Günter

19 *Lasers and Synergetics*
Editors: R. Graham and A. Wunderlin

20 *Primary Processes in Photobiology*
Editor: T. Kobayashi

21 *Physics of Amphiphilic Layers*
Editors: J. Meunier, D. Langevin, and N. Boccara

22 *Semiconductor Interfaces: Formation and Properties*
Editors: G. Le Lay, J. Derrien, and N. Boccara

23 *Magnetic Excitations and Fluctuations II*
Editors: U. Balucani, S. W. Lovesey, M. G. Rasetti, and V. Tognetti

24 *Recent Topics in Theoretical Physics*
Editor: H. Takayama

25 *Excitons in Confined Systems*
Editors: R. Del Sole, A. D'Andrea, and A. Lapiccirella

26 *The Elementary Structure of Matter*
Editors: J.-M. Richard, E. Aslanides, and N. Boccara

27 *Competing Interactions and Microstructures: Statics and Dynamics*
Editors: R. LeSar, A. Bishop, and R. Heffner

28 *Anderson Localization*
Editors: T. Ando and H. Fukuyama

29 *Polymer Motion in Dense Systems*
Editors: D. Richter and T. Springer

30 *Short-Wavelength Lasers and Their Applications*
Editor: C. Yamanaka

31 *Quantum String Theory*
Editors: N. Kawamoto and T. Kugo

32 *Universalities in Condensed-Matter*
Editors: R. Jullien, L. Peliti, R. Rammal, and N. Boccara

33 *Computer Simulation Studies in Condensed-Matter Physics: Recent Developments*
Editors: D. P. Landau, K. K. Mon, and H.-B. Schüttler

34 *Amorphous and Crystalline Silicon Carbide and Related Materials*
Editors: G. L. Harris and C. Y.-W. Yang

35 *Polycrystalline Semiconductors: Grain Boundaries and Interfaces*
Editors: H. J. Möller, H. P. Strunk, and J. H. Werner

36 *Nonlinear Optics of Organics and Semiconductors*
Editor: T. Kobayashi

37 *Dynamics of Disordered Materials*
Editors: D. Richter, A. J. Dianoux, W. Petry, and J. Teixeira

38 *Electroluminescence*
Editors: S. Shionoya and H. Kobayashi

39 *Disorder and Nonlinearity*
Editors: A. R. Bishop, D. K. Campbell, and S. Pnevmatikos

40 *Static and Dynamic Properties of Liquids*
Editors: M. Davidović and A. K. Soper

41 *Quantum Optics V*
Editors: J. D. Harvey and D. F. Walls

42 *Molecular Basis of Polymer Networks*
Editors: A. Baumgärtner and C. E. Picot

43 *Amorphous and Crystalline Silicon Carbide II: Recent Developments*
Editors: M. M. Rahman, C. Y.-W. Yang, and G. L. Harris

44 *Optical Fiber Sensors*
Editors: H. J. Arditty, J. P. Dakin, and R. Th. Kersten

45 *Computer Simulation Studies in Condensed-Matter Physics II: New Directions*
Editors: D. P. Landau, K. K. Mon, and H.-B. Schüttler

MIX
Papier aus verantwortungsvollen Quellen
Paper from responsible sources
FSC® C105338

If you have any concerns about our products,
you can contact us on
ProductSafety@springernature.com

In case Publisher is established outside the EU,
the EU authorized representative is:
Springer Nature Customer Service Center GmbH
Europaplatz 3, 69115 Heidelberg, Germany

Printed by Libri Plureos GmbH
in Hamburg, Germany